EXPLOSIONS

OF

STEAM BOILERS:

HOW THEY ARE CAUSED, AND HOW THEY MAY BE PREVENTED.

By J. R. ROBINSON,

STEAM-ENGINEER.

Merchant Books

1870

CONTENTS.

PREFACE.

WHILE it is true that the condition of many boilers now in use is such that it is a matter of surprise that so few boiler explosions occur, having their origin in excessive pressure, overheating of the surfaces above the water, in defects of materials of construction, and in the presence of scale and sediment, it is also true that there have been so many explosions not attributable to either of these causes, as to point unmistakably to the existence and operation of a power not indicated by the pressure gauge. Many theories have been advanced to account for these explosions, but without such a clearing up of the mystery as has resulted in the general adoption of means for their prevention. Many attempts have been made to construct boilers of parts so small that the explosion of one of them will not lead to very serious results ; but such boilers are so deficient in area of water-level, and in steam room, and there are such difficulties in the way of maintaining the water at the same level in the different sections — with the necessary exposure of the surfaces in order to vaporize the water thrown up mechanically mixed with the steam — that there must be so much surface exposed to the action of the fire above the water, as to cause a dangerous superheating of the steam when but little water is thrown up, or the quantity of water so thrown up will at times be so great as to unfit the boiler for use. And there are comparatively few waters so free from impurities, that such boilers will not have their surfaces so coated with non-conducting substances as to cause a rapid loss of their power. So it is found that the remedy for boiler explosions does not lie in this direction, but rather in such an improvement in the construction, and in the employment of such means for the safety of boilers having the necessary area of water-level, steam room, provision for the circulation of the water, and for reaching all the surfaces for the removal of scale and sediment, that the water level can be maintained without serious fluctuations; and that so little water will be thrown up mechanically mixed with the steam, that the action of the heat can be confined entirely to the parts of the boiler always covered by water. And notwithstanding the desirableness of avoiding the use of very large shells for high pressures because of the objections to the use of very thick plates, it is undoubtedly better to use iron of such a quality that the requisite strength can be got in large shells, than to resort to the use of sets or " batteries " of boilers of smaller diameter fed together, because of the dangers connected with the variation of the water-level in the different boilers in a set or " battery." Probably a large proportion of all the explosions on our western waters, have had their origin in this cause.

It is my aim to set forth in the following pages, in as few words as possible, — with the necessary citation of acknowledged authorities, and reports of cases in the several classes of explosions, — the causes of the explosion of steam-boilers, and the means to be used for their prevention, that all interested in these matters may be in possession of the important facts in relation to them. I do this with the feeling that a report of some of the more important facts observed by me during my experience and experiments, will be of value, inasmuch as they throw light upon the classes of explosion hitherto involved in so much mystery; and especially that a report of the wonderful manifestations of the power of the repulsive action of heat upon water, and of the explosive force of the sudden vaporization of water on the bottom of a steam-boiler, caused by the reduction of the temperature of the surfaces, as witnessed by me during recent experiments for determining the strength of certain metals when exposed to pressure and to the action of the heat in the bottom of a steam-boiler, will not only prove of interest, as throwing light upon the cause of one class of these explosions, but will be of great value as pointing to the means by which they may be prevented. These experiments show, in relation to the repulsive power of heat, as will be seen in connection with the class of explosions caused by the sudden vaporization of water on the bottom of a boiler, that it is a power always active during the vaporization of water under pressure, and ever ready to assert itself by driving the water from the surfaces exposed to the action of heat; that pressure does not appear to change the temperature of maximum vaporization; and that, while pressure alone does not change the temperature of perfect repulsion, the combined action of pressure, and of a forcible circulation of the water against the surfaces exposed to the heat, does so far overcome the power of the repulsive action, that, so long as the circulation is kept up, the temperature of perfect repulsion is practically raised. These experiments also show how a strong steam boiler may be caused to explode at, or below, the ordinary working pressure, without a sign of trouble with the water noticeable at the surface up to the time of the explosion, and without an elevation of the temperature of the boiler that it would be possible to detect by the most careful examination afterwards.

J. R. ROBINSON.

BOSTON, July 29, 1870.

STEAM-BOILER EXPLOSIONS.

Before considering the subject of Steam-Boiler Explosions, it is to be remarked, that the number of boilers so injured from various causes, as to require repairs, is vastly greater than the number of those that explode. For instance, a great many cylindrical boilers, having the fire under the shell, and the feed-water introduced on the bottom, have been broken on the bottom by the contraction of the iron in this way. The feed-water, especially when introduced at a low temperature, and fast, cools off the bottom of the boiler; but the ends of the outside course, at the "roundabout" seams, do not have their temperature reduced so fast as the other parts of the plate, because the ends of the inside course, and the non-conducting sediment between the plates, especially when the outside course is too large, come between them and the water. It will be seen that the temperature of this part of the plate, with an intense fire, is always much above that of the plate where there is but one thickness of iron; so that there may be, in feeding, a difference of temperature of 200° or more in the same plate, and within a very short distance, so that

2

every time the feed is introduced, the iron in the outside course, just at the end of the inside course, is
subjected to a strain, while, at the same time the reduction of the temperature of the bottom of the boiler
brings a strain in the opposite direction, till in time the
strength of the iron is so reduced that it gives way at a
time, when, perhaps, the temperature of the feed-water
is lower than usual, or its quantity is greater, or the
fire is a little lower ; the pressure of the steam upon the
boiler having nothing directly to do with the breaking,
excepting as a higher pressure, with its higher temperature, would make more difference in temperature
between the top and bottom of the boiler. In some
instances these breakages are so large as to amount
to over one-sixth of the circumference of the boiler,
without leading directly to any other ill effect than
letting the water out of the boiler.

Many more boilers are burned so as to need repairs,
than are thrown out of their places, because of an accumulation of scale or sediment over or around their
fires.

Many " drop flue " boilers have been broken without
leading to an explosion, by the strain, together with
the corrosion, induced in all cases — so far as I am
aware, unless in cases where the iron is protected by
scale — where steam is made in the upper part of a
boiler, while the water in the bottom is at a low and
varying temperature. Several boilers of this class that
came under my observation, — boilers about twenty-
five feet long, five feet in diameter, and running under
a steam pressure of sixty pounds per square inch, fed

on the bottom with water at times as low as 40° F., and with so little heat left in the gases when they came to the bottom of the shell, that the water on the bottom was *hardly ever* up to the temperature due the pressure, — were corroded very fast at and near the bottom, and had many leaks and breakages, but no explosion.

Many boilers have been broken, where explosion has not resulted, by the strain thrown upon the iron, because of the exposure of a part of the shell, not covered by the water, to the action of the heated gases. And many boilers have been caused to leak, without leading to explosion, because of overheating resulting from low water; the overheating causing the boiler to leak so much, perhaps, as to prevent the explosion, before the tensile strength of the iron was so reduced, by the elevation of temperature, as to lead to the rupture of the boiler by pressure.

Many boilers are injured, so as to need repairs, by a repulsion of the water from their surfaces exposed to the fire (below the surface of the water), when, either because of the presence of a scale or coating that is not removed or changed by the elevation of temperature, or when, from some other cause, the reduction of the temperature of the overheated surfaces is not so rapid as to lead to explosion.

And this leads to the consideration of the matter of explosions caused by low water. While it is undoubtedly true, that an explosion not preceded by a sudden increase of pressure will result from low water, whenever the strength of the iron of the boiler, by its elevation of temperature, is reduced below the point required to withstand the steam pressure, yet, in view of the fact demonstrated by the experiments of the Committee of the Franklin Institute,* and by the experiments of Mr. Fairbairn,† that the tensile strength of most boiler iron is increased by an elevation of temperature, so that it is greater at about 400° F. than at any temperature below that, and that its tensile strength is not seriously reduced at 600° F. (in fact, some iron proving stronger at 600° than at any lower temperature), and that thus, for some time before the reduction of the tensile strength of the iron, the boiler is in a condition to vaporize suddenly a greater or less quantity of water, probably most explosions from low water have been preceded by a sudden elevation of pressure, not necessarily great; but so sudden as to rupture the boiler, and to let so much steam escape that the reduction of pressure shall cause the water in the boiler to be thrown, in the act of giving off its steam, with such force as to shatter the boiler. That the violence of boiler explosions is due very largely to the percussive action of the steam and water contained in them at the time of rupture, there can be no doubt. For a very clear setting forth of this action,

---

* For a full report of the very interesting experiments of this Committee, on the causes of boiler explosions, and for several able papers relating to the subject, see Volumes 17, 18, 19, and 20, New Series, "Journal of the Franklin Institute."

† For a full report of the experiments of Mr. Fairbairn, which are of great value to all interested in the strength of boilers, see " Useful Information for Engineers." London : Longman, Green, Longman, & Roberts. First Series, 1856 ; and Second Series, 1860.

see " Steam Boiler Explosions," by Zerah Colburn ; London: John Weale, 1860. Mr. Colburn says: " But the momentum of the combined steam and water discharged, as Mr. Clark has suggested in his communication already referred to, would probably be sufficient to overcome the resistance of the material of the boiler, and to rend it open, not only along seams of rivets, but, as is often the case, through solid iron of the strongest quality."

Mr. D. K. Clark, in the communication to which Mr. Colburn refers, and which is given by him in full, says: " And I beg leave to suggest, that the sudden dispersion and projection of the water in the boiler against the bounding surfaces of the boiler is the great cause of the violence of the results; the dispersion being caused by the momentary generation of steam throughout the mass of the water, and its efforts to escape. It carries the water before it, and the combined momentum of the steam and water carries them, like shot, through and amongst the bounding surfaces, and deforms or shatters them in a manner not to·be accounted for by simple overpressure, or by the simple momentum of steam." Mr. Colburn, in giving a summary of this action, says: " The distinct and consecutive operations into which a boiler explosion, although practically instantaneous, may probably be resolved, are these : —

" First. The rupture, under hardly, if any, more than the ordinary working pressure, of a defective portion of the shell of the boiler, — a portion not much, if at all, below the water-line.

" Second. The escape of the free steam from the steam-chamber, and the consequent removal of a considerable part of the pressure upon the water, before its contained heat can overcome its inertia, and permit the disengagement of additional steam.

" Third. The projection of steam, combined — as it necessarily must be — with the water, with great velocity, and through a greater or less space, upon the upper sides of the shell of the boiler, which is thus forced completely open, and perhaps broken in pieces.

" Fourth. The sudden disengagement of a large quantity of steam from the heated water, now no longer confined within the boiler, and the consequent projection of the already separated parts of the boiler to a greater or less distance."

To return to the consideration of the subject of the sudden elevation of pressure when the water is low, — it is to be borne in mind that at such a time all the parts of the boiler above the water, together with the steam contained in it, is at a temperature above that due the pressure within the boiler; so that, in case any of the water in the boiler, having a temperature due the pressure, be thrown up, *all* the heat imparted to it is expended in vaporizing the water, and as a consequence in an elevation of pressure. That this rise of pressure may be so violent as to lead to the rupture of the boiler was pointed out by Mr. Jacob Perkins, several years ago. Mr. Perkins, who had had great experience in the use of steam of high pressure, says, in speaking of the causes of boiler explosions (see " Journal of Franklin Institute,"

New Series, Vol. 17, p. 371) : " The second cause of explosion, which I some years since accidentally discovered and published, — and which explanation has since been experimentally proved to be correct by the celebrated French philosopher, M. Arago, — arises from the water getting too low in the boiler. The fire then impinging on that part of the boiler which is above the water causes the heat to be taken up by the steam, which rises by its superior levity to the top of the boiler, causing it sometimes to become red-hot, and so elevating the steam to a much higher temperature than its pressure would indicate. Now when the boiler is in this state, and the safety-valve suddenly raised, the water will be relieved from steam pressure and rush up amongst the surcharged steam, which thus receives its proper dose of water ; at the same time, that part of the boiler which has been raised in temperature, giving off its heat to the water so elevated, steam is generated in an instant of such force as no boiler hitherto made can resist. This kind of explosion has of late been very frequent and disastrous, particularly in America."

The correctness of this theory of Mr. Perkins has been questioned, particularly in relation to the effect of superheated steam, because of the fact of its low specific heat ; but it is to be borne in mind that the action of the superheated steam upon the water thrown up mechanically mixed with the steam, while it does not vaporize a large quantity of it, acts so rapidly, — so explosively, so to speak, — as to so completely disperse the remainder over every part of the overheated surface as that the rapidity of the elevation of pressure shall be very much greater than it would be but for the presence of superheated steam; and also that the boiler at such a time, at and near the water-level, is under such a strain from unequal expansion, that the shock produced by the sudden vaporization of even a small quantity of water might break it, even without such an overheating above the water-line as would seriously reduce the tensile strength of the iron.

The experiments of the Committee of the Franklin Institute are conclusive, as to the fact of the sudden elevation of pressure when water is thrown into an overheated boiler. In relation to this matter, the Committee say: "It has been supposed that because the metal of a boiler was heated above the temperature at which the metal would produce steam most rapidly, it was impossible to account for the formation of quantities of highly elastic steam by such a cause. The Committee determined to make the fact of the production of high steam by intensely heated metals the subject of a direct experiment, and under circumstances as nearly similar as possible to those which may occur in a boiler of which some parts, as the sides or interior flues, may become unduly heated when not in contact with water."

In these experiments the temperature of the water injected was 70°. The pressure on the boiler at the instant of the commencement of each experiment was one atmosphere, the temperature of the steam within the boiler, as shown by a thermometer near the bottom, varying from 306° to 448°. Regarding the working of the injected water, under the repulsive action of the overheated boiler, the Committee, after explaining how the water was injected, — at the back end of the boiler, — and that its course " could be distinctly marked after the bottom of the boiler had been heated to redness, and was examined through the glass window," say: "The force of the pump carried it to the front end, nearly; the boiler being slightly inclined to the back end, the water slid back in one or more dark masses, moving down the central line, or diverted up the sides, greatly agitated, and frequently changing its shape. The water generally disappeared at the back end, though parts were retained by accidental spots of sediment, and disappeared upon them." In every instance the injection of water was followed by a rapid rise in pressure; and at last, when the temperature of the steam near the bottom of the boiler was 448°, and when the metal of the bottom of the boiler was red, the injection of ten fluid ounces of water at 70° was followed by a sudden rise of pressure to twelve atmospheres, and by the shattering of one of the plate-glass windows of the boiler. The Committee say: " In the last experiment the glass window, of the fire end of the boiler, blew out with a quick, sharp report, as loud as that of a musket; the fragments of glass from a hole in the centre of the plate were projected through a window, about three feet from the boiler, and could not be found. The number of twelve atmospheres is placed opposite to this experiment, as being an approximate result. In the act of observing the gauge, the glass burst, and the mercury at once fell: the number of inches at which the mercury had certainly risen, and above which it was, — by an undetermined quantity, not, however, very considerable, — was noted ; and from this the pressure given in the table is calculated. Here explosive steam was generated by the injection of water upon red-hot iron, and in a time not exceeding one or two minutes at the most, the interval between the last stroke of the pump and the explosion not having been sufficient to note the height of the gauge."

In relation to the fact that there was not water enough injected in the experiments to give the greatest rise of pressure, the Committee say : " By comparing the temperature of the steam in these experiments, with its observed pressure, it will be seen that not in one of them was water enough injected to give the steam a density even approaching that corresponding to its temperature : for example, 336° F." — the temperature of the steam, as shown by a thermometer near the top of the boiler, after the injection of the water — " should give a pressure of $7\frac{3}{4}$ atmospheres, instead of 3·3 the observed pressure ; 338° should give more than 14 atmospheres (Arago and Dulong), instead of 8·2 the observed pressure ; and 448° about $27\frac{1}{2}$ atmospheres, instead of· 10. The violence of the effect was not, therefore, carried so far as it might have been to produce the greatest effect ; and yet within two minutes the pressure was raised from one to twelve atmospheres." In relation to the fact that the temperature of their boiler was too high for the most rapid vaporization, in these experiments, the Committee say : " Though it has been shown that water thrown upon red-hot metal is adequate to produce explosive steam, even when it does not cool the metal down to the temperature of most rapid vaporization, it is not the less true that metal more than two hundred degrees below a red heat, in the dark, is in the condition to produce even a more rapid vaporization of the water thrown upon it, than at a red heat."

It will be noticed that the conditions in these experiments are unlike those in a steam boiler in which the water is low, and water is thrown up by a sudden reduction of pressure, in this: that the water, by the repulsive action of the overheated iron, is thrown up more in masses, and consequently that it would not be brought into such an intimate admixture with the superheated steam, nor thrown so completely over all the overheated metal of the boiler, as in that case; and that the effect of the superheated steam would be less than in a case where its pressure was greater, — the temperature being the same; — so there is every reason to suppose that the elevation of pressure would be more rapid still, in case of low water and the same overheating above the water, than was the case in these experiments.

In order to test the question of the action of the superheated steam alone in raising the pressure, in such a case, and also with a view to test the correctness of that part of the theory of Mr. Perkins, the Committee made some experiments, that appeared to demonstrate that no rise of pressure would be caused by injecting water into superheated steam, and that thus the correctness of the theory of Mr. Perkins was disproved; but the conditions in the experiments were unlike those in the case supposed by Mr. Perkins in this: that the steam was not superheated from below, but by a fire on the top of the boiler; * that the water injected was not at a temperature due the pressure within the boiler, but below 212°; and that in injecting the water, it was not brought into such contact with the superheated steam as to affect it, or to be affected by it; for the report of these experiments shows that, in every instance, the thermometer in the steam in the top of the boiler, stood as high, or higher, after as before the injection of the water. It may be difficult to decide just where the line is to be drawn between the effect of the superheated steam and of the over-heated metal, in producing a dangerous rise of pressure in an overheated boiler, in which there is a throwing up of the water by a sudden reduction of pressure, caused either by the raising of a safety-valve, the starting of an engine, or by the sudden opening of any valve connected with it; but of the fact of the rise in pressure, and that it has ruptured, and so caused the explosion of many boilers, there can be no doubt.

Explosions of boilers from a steady increase of pressure, without overheating, occur whenever the pressure

---

* In relation to this, Mr. Perkins, in a communication to the "Journal of the Franklin Institute," Vol. 20, p. 84, after stating how he got the results indicated in his theory, says: "If I had made the fire on the top of the boiler, as the Committee of the Franklin Institute did in their experiments, I should have made the same mistake; and instead of surcharging the mass of the steam, I should only have surcharged a small film next the heated metal, and have left the rest perfectly saturated with water, and quite unfit for receiving a part of that fluid which would only serve to lower the temperature and pressure, which was shown in their experiments."

of the steam is increased beyond the strength of the boiler; and the rupture, or break, will commence at the weakest part of the boiler. The violence of the explosion will depend upon the pressure at the time, and upon the point, as regards the water-level, at which the boiler yields.

Such explosions, when they occur in cases of properly constructed boilers, can only be produced by pressures very much above the ordinary working pressures, and may be produced by an accidental or a wilful overloading of the safety-valve; by the adhesion of the valve to its seat; by the closing of a stop-valve between the safety-valve and the boiler; by the failure of the valve because of the working loose of its seat, so that when the valve rises enough to make a strong upward current of steam, the seat is taken up by it, and the opening closed; or by the closing up of the hole in the seat upon the spindle or wings of the valve, because of the greater expansion of the seat than of the case, with such force as to lead to a great over-pressure. I have seen several instances of such contraction, where the valve was held so securely as to withstand a pressure very much greater than that due the load on the valve. In one instance, that of a six-inch valve, which had been in use for some years, and perfectly free, which was taken out of the case when there was a little steam floating up through the seat, with a strong current of cold air past the case, the temperature of the seat was so much above that of the case, as to close up the hole in the seat to such an extent that the spindle would not enter it without a considerable reduction of its size.

Mr. Fairbairn supposes that a large proportion of all the explosions that occur, are produced by a continuous increase of pressure without the means of escape. He says (see his " Useful Information for Engineers ") : " So many accidents have occurred from this cause,—the defective state of the safety-valves, —that I must request attention while I enumerate a few of the most prominent cases that have come before me. In the year 1845 a tremendous explosion took place in a cotton-mill in Bolton. The boilers, three in number, were situated under the mill; and from the unequal capacity and imperfect state of the safety-valves (as they were probably fast), a terrific explosion of the weakest boiler took place, which tore up the plates along the bottom, and, the steam having no outlet at the top, not only burst out the end next the furnace, demolishing the building in that direction, but tore up the top on the opposite side, and the boiler was projected upwards in an oblique direction, carrying the floors, walls, and every other obstacle, before it ; ultimately, it lodged itself across the railway, at some distance from the building. Looking at the disastrous consequences of this accident, and the number of persons (from sixteen to eighteen) who lost their lives on the occasion, it became a subject of deep interest to the community that a close investigation should be immediately instituted ; and a recommendation followed, that every precaution should be used in the construction as well as the management of boilers.

" The next fatal occurrence on record in this district was at Ashton-under-Lyne, where a boiler exploded under similar circumstances, namely, from excessive interior pressure, where four or five lives were lost; and again, at Hyde, a similar accident occurred from the same cause, which was afterwards traced to the insane act of the stoker or engineer, who prevented all means for the steam to escape by tying down the safety-valve."

Mr. Fairbairn gives the facts in relation to the "terrific explosion at Rochdale, accompanied with great loss of life," of a boiler, the ordinary working pressure of which was fifty pounds to sixty pounds per square inch. He says: "With the exception of some parts of the boiler and fragmental parts of the machinery, which had been removed when searching for the bodies of those killed, I found the buildings, steam-engine, boiler, and machinery, a heap of ruins. The boiler was torn into eight or ten pieces; one portion (the cylindrical part) flattened and embedded at a considerable depth in the rubbish, and the two hemispherical ends burst asunder and driven in opposite directions to a distance of thirty to thirty-five feet from the original seating of the boiler. Other parts of the cylinder and ends were projected over the buildings across Gas-house Lane, and lodged in a field at a distance of ninety yards from the point of projection. To one of these parts was attached the 2″ safety-valve, which was torn from the boiler by the force of the explosion, and carried, along with its seating, over a rising ground to a distance of nearly two hundred and fifty yards. The other portion of the cylindrical part of the boiler was found on the opposite side in the bed of the river; and the hemispherical end of this part (furthest from the furnace) was rent in two, and thrown on each side to a distance of thirty or thirty-five feet. These two pieces had evidently come in contact with the chimney, razed it to the ground, and finally lodged themselves in the margin of the river."

After a careful examination of the parts of the exploded boiler, Mr. Fairbairn says, in relation to its strength: "In the question before us I find the boiler with hemispherical ends, 18 feet long, 5 feet diameter, and composed of plates $\frac{5}{16}$ of an inch thick, to be equal in its powers of resistance to a pressure of three hundred and thirty-five pounds on the square inch; but finding one of the plates under $\frac{5}{16}$ in thickness, I have reduced its power to three hundred pounds, which I consider the force at which it would burst."

Mr. Fairbairn also gives the facts in relation to the locomotive "Irk," "which in February, 1845, blew up and killed the driver, stoker, and another person who was standing near the spot at the time. A great difference of opinion as to the cause of this accident was prevalent in the minds of those who witnessed the explosion, some attributing it to a crack in the copper fire-box, and others to the weakness of the stays over the top. Neither of these opinions was, however, correct, as it was afterwards demonstrated that the material was not only entirely free from cracks and flaws, but the stays were proved sufficient to resist a pressure of one hundred and fifty to two hundred pounds on the square inch. The true cause was afterwards ascertained to arise from the fastening down of the safety-valve of the engine (an active fire being in operation under the boiler at the time), which was under the shed, with the steam up, ready to start with the early morning train. The effect of this was the forcing down of the top of the copper fire-box upon the blazing embers of the furnace, which, acting upon the principle of the rocket, elevated the boiler and engine of twenty tons weight to a height of thirty feet, which, in its ascent, made a summersault in the air, passed through the roof of the shed, and ultimately landed at a distance of sixty yards from its original position."

Mr. Fairbairn also reports the case of the explosion of a locomotive on the Eastern Division of the London and North-Western Railway, and in connection with it a full account of the very interesting experiments in relation to its probable strength. In respect to the appearance of the boiler after the explosion, he says: "I found one side of the fire-box completely severed from the body of the boiler, the interior copper box forced inwards upon the furnace ; and, with the exception of the cylindrical shell which covers the tubes, the whole of the engine was a complete wreck." This engine was about thirteen years old, had run 104,723 miles, and had been worked under a pressure of sixty pounds per square inch. The inside fire-box (copper), which was originally $\frac{7}{16}$ of an inch thick, had been reduced by wear so that it was but little over $\frac{3}{8}$ of an inch, and was "perfectly free from flaw or patch." The outer shell

was also good, and " nearly of the original thickness."
The screw stays were originally $\frac{11}{16}$ of an inch in diameter, but were reduced from oxidation.   The report
says : " With the exception of one stay, which was on
the top row, the one most reduced from oxidation was
half an inch in diameter."   They were placed $5\frac{3}{8}$
inches by 5 inches from centre to centre.

Owing to a difference of opinion regarding the
strength of the boiler, it appearing to be improbable
that the steam pressure could have increased from
sixty pounds (the pressure at which it was blowing off
when the safety-valve was screwed down) to a pressure
so great as would cause the boiler to yield, taking Mr.
Fairbairn's estimate of its strength to be correct, in
the time (twenty-five minutes) between the screwing
down of the valve and the explosion, experiments were
made to test the matter.   Regarding the strength of
the sides of the fire-box, Mr. Fairbairn says : " Taking into account the tensile strength of the stays — in
their corroded state — of the side of the fire-box, which
to appearance was the first to give way, I find that a
force of three hundred and eighty pounds upon the
square inch would be required to effect rupture." . . .
" Assuming, therefore, that the ends of the screws
were riveted, and sound in other respects, we may rea-
sonably conclude that a strain of not less than four
hundred and fifty to five hundred pounds upon the
square inch would be required to strip the screws or
tear the stays themselves asunder."   To test this matter still further, Mr. Fairbairn had a new box made with
$\frac{1}{2}$ inch copper on one side, and $\frac{3}{8}$ inch iron on the other,

stayed with $\frac{3}{4}$ inch screw stays, well fitted and riveted over, and placed five inches from centre to centre ; and this stood a pressure of seven hundred and eighty-five pounds to the square inch, with a bulging of the sides of less than $\frac{1}{10}$ inch, and yielded by drawing the head of one of the stays through the copper, at eight hundred and fifteen pounds per square inch. Regarding this, he says : " The above experiments are at once conclusive as to the superior strength of the flat surfaces of a locomotive fire-box, as compared with the top, or even the cylindrical part, of the boiler."

The boiler of an engine exactly like the one that exploded, by the same builders, of the same age, and that had run about the same number of miles, was tested ; and one of the " bolts of the cross-bar over the fire-box broke," at a pressure of $207\frac{1}{2}$ pounds per square inch ; but the appearance of the exploded boiler was such, that Mr. Fairbairn concluded that its crown sheet did not yield first, and that it " could not have burst under a pressure of less than three hundred to three hundred and fifty pounds upon the square inch."

In all these cases of explosion caused by a steady increase of pressure, reported by Mr. Fairbairn, it is seen that his knowledge of the strength of materials leads him to the opinion that they were caused by very high pressures ; and it is undoubtedly true that quite a large proportion of the explosions, caused by a steady increase of pressure without overheating, take place at pressures very much above those at which the boilers are designed to work.

As bearing upon this part of the subject of the strength of boilers to withstand a steady pressure, it will not be out of place to mention the case of an old boiler which I tested several years ago. This was forty-two inches in diameter, about twenty-eight feet long, with two twelve-inch flues through it. The heads were of cast iron; the iron in the shell appeared to have been originally $\frac{5}{16}$ of an inch thick, and that in the flues $\frac{3}{16}$ of an inch. It had been in use over twenty years, and it was known that the water had been so low in it, at one time, as to lead to the supposition that it might have been seriously injured by the over-heating; it had also a heavy indentation on its lower side, caused by a settling of the front, thereby bringing too much weight upon a little pier on the bridge wall; yet it stood a pressure of two hundred and forty pounds per square inch without a sign of yielding.

Notwithstanding this great strength of good boilers, it is undoubtedly true that there have been explosions of boilers at their ordinary working pressures, without overheating, and without sudden increase of pressure, because of defects in material, design, or workmanship. As, for example, a plate that has to be flanged may be so poor, or worked with so little skill, as to have so little strength left in the bend, that a little alteration of form, from variation of pressure, on account of the defective staying of the flat part of the plate, may cause a break, that, above the water-level, may lead to explosion.

A boiler may be so poorly stayed that a very little corrosion, around an imperfect weld in one of the stays, may lead to explosion. Or a plate may have its strength so reduced, at and near a longitudinal seam, in punching, bending, pinning, and riveting, that the little alteration of form that takes place at this point at every variation of pressure, may cause a break, particularly in a wide plate, that will lead to an explosion; or a leak from such a source may, by corrosion, so reduce the strength of the iron at that point, that, in combination with the alteration of form, an explosion may ensue.

Explosions have been caused by blisters, where the imperfect weld in the plate has been so large, and so near the side next the water, as to leave so little strength, after the side of the plate next the fire has parted off, with the consequent loss of its strength because of elevation of temperature, as to be ruptured by the ordinary working pressure.

Explosions have occurred from overheating caused by the presence of scale and sediment. They may be caused by the overheating of the bottom of the boiler by the presence of scale, that adheres to the iron over a limited surface, till its power of resistance is reduced below that of the ordinary working pressure of the boiler, so that a rupture occurs, large enough, perhaps, to merely throw the boiler out of place, or, it may be, extends so as to cause a violent explosion; or the explosion may be caused by an overheating, not so high as to seriously reduce the tensile strength of the iron, but of so much of the surface of the bottom of the boiler, as, by the strain from unequal expansion, to cause a break through a "round about" seam; or by a scale which adheres till the boiler becomes so much over-heated where the heat is most intense, as that the expansion of the iron shall detach scale enough to cause the breaking of the bottom of the boiler, by the combined action of the strain caused by the contraction of the uncovered part of the boiler, and the local pressure of the steam suddenly generated under water ; or it may be caused by the detachment of a considerable surface of scale from the bottom of the boiler, when the temperature of the uncovered part is but very little above that of maximum vaporization, which may cause the breaking of its top by the blow from the water put in motion by the sudden generation of steam on the bottom.

A boiler may be caused to explode by an overheating caused by the presence of loose, heavy scale, that, after accumulating on the surfaces of the boiler, be-comes detached, and accumulates over the fire. In several instances I have known this scale to become detached from boilers just after they had been cleaned out, when it was supposed that all was removed that it was possible to get off, and to be deposited in a heap exactly over the hottest part of the fire, leaving the other parts of the boiler as clean as if they had just been swept. This loose scale does not adhere to the boiler so as to become detached by its expansion, but follows down on the iron till the rupture occurs.

Regarding scale and deposits of sediment, the Com-mittee of the Franklin Institute say : " The undue heating of parts of a boiler may be produced by de-posits. No cause of undue heating is better made out than this : the least that can happen after the accumulation of sediment is the injury of the boiler, — perhaps its bursting, — and a true explosion may result. Two violent explosions — at Bowen's Mill, and at McMickle's Mill, in Pittsburg — are fairly attributable to the effect of sediment; and there does not appear, in either case, to have been a deficiency of water. M. Arago mentions an instance of a rent made in a boiler, at Paris, by the accidental resting of a rag on the bottom of the boiler."

We now come to the consideration of those explosions, of which there have been so many, occurring when it was known that the pressure was not above the ordinary working pressure, and that the water was not low; when there was no *appearance* of overheating on any part of the boiler, when examined after the explosion, and when, from the most careful examination, there were no indications of weakness that could account for the explosion; in fact, in many instances, the indications being that the boiler, a moment before the explosion, was very much stronger than many boilers running without a leak, or sign of failure, under the same pressure of steam; and in some instances parts of the exploded boiler itself, that, to all appearance, were very much weaker than the ruptured parts, were not only not broken, but showed no sign of having been subjected to a severe pressure, even where the top of the shell was completely shattered.

In relation to these cases, Mr. Colburn says: " Yet those who have given any attention to the subject of boiler explosions are aware that they frequently occur

when, without any overheating of the plates, the
pressure stood, but a moment before, at the ordinary
working point. In the case of the locomotive boiler
which exploded in the summer of 1858, at Messrs.
Sharp, Stewart, and Co.'s, at Manchester, the pressure,
as observed upon two spring-balances and a pressure-
gauge, stood at 117 pounds to 118 pounds a minute
before the explosion, both valves blowing off freely at
the time. The part of the boiler which exploded was
the ring of plates next the smoke-box, out of the
influence of any part of the fire. The fact, therefore,
of the violent explosion of a strongly made boiler at
117 pounds is a proof that it is not necessary to
assume and to account for the existence of any pressure above that point. On the 5th of May, 1851, a
locomotive engine, only just finished, burst its boiler
in the workshop of Messrs. Rogers, Ketchum, and
Grosvenor, at Paterson, U.S. I was on the spot but
a few moments afterwards, and found the effects of the
explosion to be of the most frightful character: a considerable portion of the three-story workshop being
blown down, whilst four men were instantly killed,
and a number of others were injured, one of whom
died soon afterwards. Several of the men, who,
although immediately about the engine at the time,
escaped unhurt, unanimously declared that the safety-
valves were blowing off before the explosion, and that
the two spring-balances indicated, but a moment before
the crash, a pressure of but 110 pounds per square
inch."

These are cases of the explosion of boilers at pressures that, without proof to the contrary, may be taken to be very much below their powers to resist pressure unaccompanied by shocks. In one instance of explosion that I investigated in 1854, there was no appearance of overheating on any part of the boiler, or evidence that the pressure was over 80 pounds per square inch a moment before the explosion. The top of the shell was broken up so as to lead to the supposition that the iron must be very poor; but on testing a strip cut from one of these small, brittle-looking pieces, that had been blown out clean, I found it to be stronger than new Low Moor, or Bowling iron, tested at the same time. The new Bowling iron broke at 55.632 pounds per square inch; the new Low Moor, at 56.704 pounds; and that from the exploded boiler, at 58.250 pounds.

And in a case that I investigated in 1856, in which the top of the shell of a locomotive was pretty much all blown off when the engine was running on the road, under a pressure of about 100 pounds per square inch, two strips of iron cut from a piece of the exploded boiler broke, one with 43.392 pounds per square inch, and the other a very little under: these strips were cut from a little piece, blown out of the plate in the top of the shell, that appeared to have been the first to yield; and its weakness would indicate that the boiler might have yielded to overpressure. But that overpressure was not the cause of the explosion, was proved by the fact that parts of the flat surfaces of the fire-box, which were so weak, because of broken screw stays, as that they must have yielded to a pressure very much below that which would have caused the rupture in the shell, — even if the unbroken stays and the plates of the fire-box were of iron of the maximum strength, — were not opened in the least. The broken stays were in the upper row of the sides of the fire-box, and it was perfectly evident they had been broken for a long time; but the sides of the fire-box and the crown sheet stood up square and good. In this case, it was noticed, in leaving the station just before the explosion, that the water was not working right. I was told, " She had so much water in her that she could hardly be got away from the station," and the fire-box ends of the copper tubes showed clearly that they had been overheated. But it was not clear how much of this overheating might have been done after the boiler exploded, particularly as it was known that there was a heavy fire at the time, which was not removed till some little time after the explosion. But from the fact that tubes which were very nearly closed up by the explosion, and through which there was no passage for the heat, had so nearly the same appearance as others that were not so closed, I am satisfied that a part at least of this overheating was done before the explosion; and that the explosion was caused by overheating below the level of the water, and around the fire. The engineer was killed, and the fireman injured; but the latter told me he saw the engineer try the water but an instant before the explosion, and it was high enough; the trouble with the water at the station had so called his attention to the fact that it was not working right, that he was watching it continually.

Mr. Colburn, in speaking of this class of explosions, says: "Locomotive boilers have burst in the plates next to the smoke-box, beyond the reach of the fire, and where the boiler is believed to be stronger than about the fire-box. As has been observed, the dome, if it open from the ring of plates in question, weakens it materially; but explosions have occasionally occurred in this part of a boiler, either having no dome, or having one only over the fire-box. A fact which was some time since communicated to me by George S. Griggs, Esq., Locomotive Superintendent of the Boston and Providence Railroad, U.S., may assist in explaining this somewhat anomalous mode of explosion. In one or two cases of locomotive boiler explosions, Mr. Griggs found, upon examination, that whilst none of the upper tubes had been burnt, others, lower down, exhibited unmistakable indications of having been smartly scorched; the solder used in brazing being more or less melted."

That the explosions now under consideration are caused by an overheating near the bottom of the boiler, causing the water to be thrown with such force as to break the top, I think there can be no doubt. Mr. Rankine (see his "Manual of the Steam Engine and other Prime Movers;" London and Glasgow: Richard Griffin & Co., 1859) says of this class of explosions: "There is much difference of opinion as to some points of detail in the manner in which this phenomenon is produced; but there can be no doubt that its primary causes are, first, the overheating of a portion of the plates of the boiler (being in most cases that portion called the *crown of the furnace*, which is directly over the fire), so that a store of heat is accumulated; and, secondly, the sudden contact of such overheated plates with water, so that the heat stored up is suddenly expended in the production of a large quantity of steam at a high pressure. Some engineers hold, that no portion of the plates can thus become overheated, unless the level of the surface of the water sinks so low as to leave that portion of the plates above it, and uncovered; others maintain, with M. Boutigny, that when a metallic surface is heated above a certain elevated temperature, water is prevented from actually touching it, either by a direct repulsion, or by a film or layer of very dense vapor; and that when this has once taken place, the plate, being left dry, may go on accumulating heat and rising in temperature for an indefinite time, until some agitation, or the introduction of cold water, shall produce contact between the water and the plate, and bring about an explosion."

Mr. Bourne, in the eighth edition of his "Treatise on the Steam Engine," London, 1868, in relation to this matter, says: "There can be no doubt that the water is sometimes repelled from the metal in the same manner as would be done if it were in the spheroidal state; and explosions have, no doubt, had their origin in this phenomenon. The water appears to be repelled from the iron in those parts where the heat is greatest."

Mr. Colburn, in speaking of the rising of water because of the condensation of the steam above the water, says, regarding the force of the blow given by it: "In this case . . . it would not be necessary to assume the existence of any defect in the boiler; for, when the water once struck violently, the soundest iron would probably be broken, and the strongest workmanship destroyed."

The experiments of the Committee of the Franklin Institute demonstrated that the temperature of maximum vaporization of a clean, rough, iron surface, — like that of a clean iron steam-boiler, — is $346\frac{1}{2}°$ F., and that the temperature of perfect repulsion for the same is 385° F.; and the temperature of maximum vaporization for iron, " highly oxidated but clean," is 381° F., and that of perfect repulsion for the same is 433°; the temperature of maximum vaporization of polished copper is 292°, while that of perfect repulsion is 315°; for the same oxidized, the temperature of maximum vaporization is 317°, while that of perfect repulsion is 338°; for copper very much oxidized and not clean, the temperature of maximum vaporization is 348°.

The Committee say, in relation to this matter: "The time of vaporization is less in the copper than in the iron, in the ratio, probably, of two to one, or nearly in the ratio of their conducting powers for heat, which are two and one half to one. . . . A repulsion between the metal and the water is perfect at from twenty to forty degrees above the point of maximum vaporization, following more closely upon the temperature of maximum vaporization in copper than in iron. At these temperatures the water does not wet the metal. . . . There can be no doubt, that at the temperatures determined

as those of maximum vaporization, an effective force
of repulsion between the heated metal and the water
has begun to be developed.

Regarding the effect of pressure upon the points of
maximum vaporization, and of repulsion, the Committee say: "It is possible, and indeed probable, that
pressure may modify these results, all of which were
obtained under atmospheric pressure. Pressure, tend-
ing to counteract the effect of the repulsion between the
heated metal and the water, would probably raise the
temperature of most rapid vaporization."

With regard to the effect of pressure in a steam-
boiler upon the point of perfect repulsion, my experiments have demonstrated that pressure, accompanied
by a rapid circulation of the water, does so far overcome the repulsive action of the heat, that practically the point of repulsion may be said to be raised
by the pressure within the boiler; but that this only
holds true so long as there is perfect circulation of the
water; and that in a steam-boiler, when the circulation
of the water is not perfect, pressure does not appear to
have any effect whatever upon the temperature of repulsion. So that a steam-boiler working under one
hundred pounds pressure, the surfaces of which are
of such a character that the temperature of perfect
repulsion at atmospheric pressure is 385°, will be liable
to just as perfect repulsion from any of its surfaces
exposed to an intense heat, whenever that temperature
is reached, unless the circulation of the water within
the boiler is so good that it shall be continually
brought into *forcible* contact with such surface; and

that the combined action of pressure and of this perfect circulation of the water is such, that water may be kept in contact with surfaces exposed to an intense heat, even up to pressures so high that the temperature of the water in the boiler is above that of perfect repulsion for the surfaces; in this case, however, my experiments show that the temperature of maximum vaporization is not changed; or if changed at all, not so changed but what it is passed, so that the power of the repulsive action is so effective as that, notwithstanding the combined action of the circulation and of the pressure, the vaporization is decreased.

In my experiments, — which were made to enable me to perfect my Safety Plugs, and to get exact proportions of the metals used for securing their conducting rods to the cases, — it was necessary that the conditions should be as nearly the same as in the case of boilers in use, as possible; and to this end, the plugs designed for the lower parts of the boilers were tested by putting them in the bottom of the experimental boiler, and causing the blast to act upon them, and upon the parts of the boiler around them. It was found, soon after the commencement of the experiments, that the variation in the results, because of the imperfect circulation of the water in the boiler, were so great, — notwithstanding the making of alterations which promised to remedy the defect, and the taking of the utmost care to keep the boiler free from any thing having a tendency to cause imperfect circulation, and also the reduction of the intensity of the heat on the conducting-rods, from which, because of the nature

of their surfaces, repulsion would take place at a lower
temperature than from the boiler, — that they were
of no value as showing the exact pressure required to
blow out the conducting-rods. I then had a boiler
made, in which the circulation of the water was so
strong, around and against the plugs, that, when the
boiler was clean, it would stand a power of blast strong
enough to raise the pressure from sixty pounds to three
hundred and sixty pounds per square inch, in from eight
and one half to nine minutes, without such a repulsion
from the conducting-rods as to overcome the combined
power of the pressure and of the circulation, so as
to seriously affect the results. I was then enabled to
complete the work of perfecting the plugs and of getting the exact proportions of the metals for them ; and
in doing this, and by experiments then made with the
first-mentioned boiler, the facts indicated above, in
relation to the effect of pressure upon the points of
maximum vaporization, and of perfect repulsion, were
brought out ; also the fact that a variation in the quality of the water, so slight as not to be apparent to the
taste, or to affect its color, as seen in a glass vessel,
did affect its circulation in a steam-boiler, in its relation to the point of perfect repulsion ; and also, that
whenever any part of the surface of a steam-boiler,
much below the surface of the water, is raised much
above the temperature of maximum vaporization, the
reduction of its temperature to that point is attended
with such a vaporization of water as to endanger the
boiler. Taking the temperature of maximum vaporization of the copper conducting-rods, in the condition

in which they were in the boiler during the experiments, to have been 320°, and their temperature of perfect repulsion to have been 340°, it will be seen that at a pressure of one hundred and forty pounds to the square inch, the temperature of the water within the boiler, and having the temperature due its pressure, is above that of repulsion, and that the conducting-rod and its surrounding case must, when the full power of the blast was on them, have been at a higher temperature than the water. The conducting-rods and cases are of pure copper, and their construction and arrangement are such as are calculated to reduce this difference in temperature to the minimum; but still, their temperature must, with the blast on, be above that of the water within the boiler, even before the point of maximum vaporization is passed. So that the fact that a conducting rod — brazed in with a metal that is known will have its strength so reduced by an elevation of temperature above 340°, that it will be broken by the pressure due that temperature, and the rod blown out — is not blown out below that pressure, when acted upon by a powerful blast, is proof that the power of the repulsive action of the heated rod and its case upon the water has not been sufficient to keep the water from the rod, so that its temperature has been elevated much above that of the water within the boiler.

This fact was demonstrated by the repeated testing of plugs above the temperature of perfect repulsion at atmospheric pressure, and as high as a pressure of three hundred and sixty pounds per square inch above the atmosphere, and consequently to a temperature nearly 100° above that of repulsion.

That the power overcoming the repulsive force in these cases was not that of pressure alone, was demonstrated in this way: after the perfecting of the plugs, and getting the exact strength of the metals, so that it was known at what pressure a conducting-rod would be broken from its case when the temperature was that due the pressure, they were tested in the first mentioned boiler, in order to learn how much their strength would be reduced by any elevation of temperature because of a repulsion of the water from the parts acted upon by the blast; and it was found that the repulsion in some instances was such that conducting-rods, so strong as to stand a pressure of three hundred and sixty pounds per square inch, with a power of blast that raised the pressure from sixty pounds to three hundred and sixty pounds, in from eight and one-half to nine minutes, would be blown out at pressures below ten pounds per square inch.

That pressure on the boiler did not have the least effect upon the power of repulsion, was shown by the fact that when the pressure was raised to fifty pounds per square inch, — at which pressure the temperature of the water within the boiler would be more than 40° below that of perfect repulsion, — by a power of blast so moderate that there was no appearance of a repulsive action, when the full power of the blast was turned on, the perfect repulsion took place just as quickly as when the pressure was but one pound per square inch, when the full blast was turned on.

In making these experiments, the heat was confined to the lower part of the boiler; so that when the repulsion was perfect, the water was so elevated into the upper part of it, out of the reach of the direct action of the blast, that the loss by radiation caused a reduction of the pressure. And in some instances the repulsive action was so strong, that the water was kept so still in the upper part of the boiler, as long as the blast was maintained, that the steam-gauge showed a steady decrease of pressure, with hardly a sign of agitation within the boiler. The return of the water to the bottom of the boiler, after the shutting off of the blast, was always attended with violent fluctuations of pressure, very much more violent than were ever produced while the repulsive action was being developed.

That the surfaces of a steam-boiler which are exposed to the most intense heat may be maintained at a temperature above that of maximum vaporization, and below that of perfect repulsion, the power of repulsion, and that of the combined action of the circulation of the water and of the pressure, so nearly balancing each other, that, so long as there is no failure in the circulation, the temperature of such surfaces will be kept below the temperature of perfect repulsion; and that, when this is the case, a reduction of the intensity of the fire is attended with a violent vaporization of water, — was demonstrated in this way. As before mentioned, I tested a great number of plugs with a strong blast, and at pressures so high that the temperature of the water within the boiler, having the temperature due the pressure, was much above the temperature of perfect repulsion for their surfaces under atmospheric pressure, without the development of such a repulsive action as to lead to an elevation of temperature so great as to seriously reduce the strength of the brazing of the conducting-rods; showing clearly that the combined action of the pressure and of the circulation did so far overcome the power of repulsion as to practically elevate the point of perfect repulsion. At the commencement of these experiments, it was the practice to turn off the blast at a pressure approaching that at which it was supposed the conducting-rod would be blown out, in order to observe its appearance; but it was found that in many instances, in a time varying from one to four or five seconds after the blast was turned off, the rod would be blown out. This occurred at pressures all the way up to over three hundred pounds per square inch above the atmosphere. Rods were blown out in this unaccountable way, as it then appeared to me; and at pressures so much below the supposed strength of the metals, as to defeat the object sought, which was the getting the exact strength of the metals at known temperatures. This practice of turning off the blast was kept up for some time after it was clear that the results were of very little value as showing the exact strength of the metals, because of a desire to subject each of quite a large lot of metals, varying very much in strength, to exactly the same treatment. The practice was then changed: the blast was not turned off after approaching the breaking point of the metals, and the experiments were repeated with much care; and then, after getting the exact strength of the

metals, it was perfectly evident that the turning off of
the blast was followed by an explosive vaporization
of water by the conducting-rod; showing that the
temperature of the conducting-rod, when the blast was
on, was above that of maximum vaporization; and
that the violence of the result was due to the reduction
of temperature to that point, and to the then explosive
vaporization of a little water before the inertia of that
above it could be overcome.

The water used in these experimental boilers was
all drawn from the same pipe, and great care was
taken to keep the boilers clean; and yet there was a
very noticeable difference in its working in the boilers,
in relation to the matter of repulsion, from day to day,
which manifested itself in the difference in power of
blast required to produce perfect repulsion, the length
of time after the blast was turned off before the return
of the water to the overheated surfaces, and in the
difference in the elevation of temperature above that
of maximum vaporization, when the point of perfect
repulsion was not reached; when all the other conditions appeared to be the same.

It thus appears that during these experiments there
were many genuine explosions, the effects of which,
owing to the great strength of the boilers and the
comparative weakness of the safety plugs, were confined to the plugs, they having broken by the first
explosive burst of steam, so that the water was not
thrown with sufficient force to break the boiler; or
in the case of a continued perfect repulsion, yielding
to the pressure at a temperature far below that at

which the strength of the boiler was seriously reduced; and that, while in the explosions of this
class (a continued perfect repulsion) there were such
indications of trouble within the boiler, that an engineer, who was familiar with the working of the
boiler, and who was watching the steam-gauge and the
water at the time, would have his attention called to
the fact, — there were in the other class (that in which
perfect repulsion was not established) no indications
whatever of trouble in the boiler till the plugs broke;
so that it would be impossible for an engineer, by any
examination he could make, to know that there was
the least trouble with the boiler up to the instant of
the explosion.

It is this class of explosions, and those to be considered farther on, caused by the overheating of the water,
that have led to so many theories of explosive gases
and mixtures. It is known to all who have had experience in the examination of exploded boilers, and also
in the testing and inspecting of old boilers that have
not exploded, that there are many boilers in use that
are defective in construction, with poor safety-valves
on them, gauge-cocks half choked up, and with large
accumulations of sediment or scale in them, so that it
is hard to understand why they have not exploded;
while, in many instances of explosion, there are no
such indications of weakness, or of defective fittings
or condition. I never saw but one exploded boiler
that bore evidence of having been in so bad condition
as very many that I have tested and inspected that
were not even suspected, before the examination, of

being in a dangerous condition. In many instances of these explosions of strong clean boilers, experts have either had to acknowledge that they did not know the cause, or to report that the water must have been low; that the safety-valve must have been fastened down; or that, notwithstanding the apparent strength of the boiler, there must have been some hidden defect in its construction or material.

Now, in order to the better understanding of the causes that produce an explosion of the class under consideration, let us take the case of a boiler in which the surfaces around the fire are clean and smooth, and very nearly uniform, and in which the circulation of the water is ordinarily good, but which has now got some element in the water that favors repulsion, or causes the water to circulate sluggishly, working with so strong a fire that the surfaces exposed to the most intense heat shall be raised to a temperature very near that of maximum vaporization, but not above it, so long as the steam pressure is maintained. Now, let the demand for steam be increased, so that there shall be such a sudden reduction of pressure that the descending current shall give off steam, and the circulation will be so broken up that the temperature of the surfaces around the fire will be raised above that of maximum vaporization: this will cause a decrease in the steaming power of the boiler, so that, without a greater demand for steam, the reduction of pressure will be more rapid, with its consequent interference with the circulation of the water, and thus the temperature of these surfaces be rapidly raised to that of perfect

repulsion. Now, let the demand for steam be so far reduced that the boiler shall make steam as fast or a little faster than it is used, and let a fire-door be opened
and a strong current of cold air be thrown upon the
overheated surfaces, their temperatures will be so
reduced that the water will return upon them and
complete the reduction to the temperature of maximum vaporization, with the consequent violent vaporization of such an amount, that the steam so generated shall throw the water above it with such force
as to break the shell of the boiler, and so cause its
explosion.

Or, without the opening of the door, let the demand
for steam be so reduced as to lead to a rapid rise in
pressure, and the interference with the descending current ceases; the circulation of the water is soon so
strong as to overcome the repulsive power; the overheated parts are reduced to the temperature of maximum vaporization, and the water is thrown by the
steam generated under it with such force as to break
the shell of the boiler, and so cause its explosion; —
without, in either case, any part of the boiler having
been overheated so as to show it after the explosion, or
so as to have reduced its tensile strength in the least.

I am of the opinion that a large proportion of the explosions of locomotives occurring while on the road, or
just after arriving at a station, are caused by their shells
being broken by water thrown from overheated plates
considerably below the surface of the water, and that
this is true of most boilers that have the upper part of
their shells blown off, without the throwing of the

lower part of the boiler. Mr. Colburn gives six cases of violent locomotive explosions, in which the engines did not leave the rails, — in two of which the fact is mentioned that they occurred just after the shutting off of the steam, — and says: " Comparatively few locomotive boilers ever leave the rails when they explode, unless the roof of the inside fire-box is crushed down." The locomotive mentioned on pages 33, 34, did not leave the rails; and the boiler mentioned on page 33 was hardly moved from its place, the tops in each of these instances being completely shattered. When explosions are caused by overpressure, boilers are usually thrown to a greater or less distance; as is shown clearly by the cases given by Mr. Fairbairn, on pages 22, 23, 24, 25, and 26.

Now, as to the force with which the water is thrown from an overheated surface. From the force of the bursts during my experiments, I am of the opinion that the return of water upon clean iron surfaces much below the surface of the water, after a repulsion, especially when the return is aided by a reduction of temperature on the fire side, will result in an instant rise to a very great pressure before the inertia of the water is overcome. I think that with four feet of water over the place, this momentary pressure might be even three hundred pounds per square inch or more, above that — beyond the reach of this local action — on the boiler at the time; and that, in all cases where there is much surface overheated at any great depth below the surface of the water, and where, either from a sudden reduction of temperature on the fire side, or from the rapid rise of pressure from the reduction of demand for steam, there is a strong return of the water to the overheated surfaces, the water is thrown with such force as to break the shell of the strongest boiler in use.

In the cases, like some of those given by Mr. Colburn, where the shells of locomotives were broken just back of the smoke-boxes, the water is evidently thrown from the tube-plate and from the sides of the fire-box at the same time, giving a forward and upward direction to the water. In the cases given on pages 33 and 34, the water was thrown against the shell, over, or just back of, the fire-boxes. In my experiments, the conducting-rod yielded to the pressure that would have thrown the water; and boilers sometimes yield in the same way, the overheated surfaces not being strong enough to hold till the inertia of the water is overcome. In one instance of this kind that came to my knowledge, the lower part of the fire-box tube-plate of a locomotive yielded to the pressure by stripping the threads and riveting of ten or twelve screw stays; but the steam which had caused the burst then escaped into the fire-box, so that the local pressure was removed; and although the strength of the parts was so reduced by the stripping of so many stays, they withstood the pressure of one hundred and twenty pounds on the boiler at the time. This engine, which was a first-class passenger engine, but two or three years old, had just stopped at a station, and the fireman had had the fire-door open with the blower on, and was just in the act of closing it, or had just closed it, when the burst occurred, and he was killed by the rush of steam from the fire-box : there had been no such marked defect in the circulation of the water in this boiler as to attract attention, nor was the over-heating sufficient to show the fact after the rupture. The stripped stays were $\frac{7}{8}$ of an inch in diameter, and there was no appearance of any imperfection in them, or in the plate forced off, or in the workmanship ; but the stays were placed $5\frac{1}{2}$ inches from centre to centre, and to this fact is due the stripping of the stays, and the saving of the boiler ; for there can be no doubt what-ever, that if these stays had been a very little nearer together, the water would have been thrown, and a first-class explosion would have been the result.

Taking the results of the experiments of Mr. Fair-bairn on the strength of such flat surfaces, stayed with well-riveted screw stays, as a guide, I am of the opin-ion that this tube-plate would not have been forced off in the way it was, at a pressure less than three or four times that on the boiler outside of this local action, at the time ; and the fact that the yielding did not extend to the destruction of the boiler, after the great reduc-tion of strength caused by the stripping of so many stays, and which would not be attended by a sensible reduction of pressure outside of the local action, while the plate was in motion, is a perfect demonstration that the plate yielded to a pressure very much greater than that on the boiler outside of the influence of this action, and that it was a local pressure that was in-stantly and entirely removed by the opening of the stay-holes.

I am confident that explosions of this class have had their origin in overheating, resulting from the surfaces around the fire having become coated with some substance or substances that favor the repulsive action, so as to lead to the overheating by a comparatively moderate fire; the coating remaining unchanged till there is a certain elevation of temperature (above that of maximum vaporization), and then being entirely removed, or its character so changed by the elevation of temperature, as to leave a surface of such a character that its temperature will be reduced so fast from the circulation of water within the boiler, without any great increase of pressure, or without any reduction of temperature on the fire side, as to lead to the throwing of the water and destruction of the boiler.

I think this is the case in most new boilers, like those of the new locomotives given by Mr. Colburn on page 32; * and it is probable that in many of the explosions of this class the overheating is assisted, more or less, by some coating of this nature. In a case of the explosion of a cylindrical boiler, with the fire under its shell, the first time it was used, after the putting in of a new plate over the furnace, it was noticed, not long after steam was got up, that the new plate was so overheated as to cause the leaking of

---

* In the case of the explosion at Paterson, mentioned by Mr. Colburn, the crown-sheet of the boiler was weaker than the shell, so that when the water thrown up from the sides of the fire-box met above it, the shock caused it to yield, so that it was forced down and the engine thrown into the air.

its seams. The opening of the doors to reduce its temperature was followed in a few minutes by the explosion of the boiler. The overheating of this plate was undoubtedly due to a coating on it that favored repulsion; and it is possible that the return of the water to it, after the change in its character, or its removal by the elevation of temperature, might have been so gradual as to have avoided the explosion, but for the sudden reduction of temperature on the fire side, caused by the opening of the fire-doors.

The circulation of water in boilers is frequently affected by changes in the character of the water in them, apparently very slight, and which have in many instances been produced by some substance which has accumulated in the water-pipes, and been washed through into the boilers, without its presence being suspected; and some water, that ordinarily circulates well, will be seriously affected by a very slight interference with the descending currents. Two boilers, about twenty-eight feet long, four feet in diameter, with two sixteen-inch flues in each, were set side by side, with one furnace under the two: they were exposed to the action of the flame and heated gases, to a point above that of the middle gauge-cock, and in some places nearly up to the upper cock, but the brick-work was but little off from the boilers from the centre up on the outsides, and they were set so near together as to make the passage between them quite narrow, so that their surfaces were not exposed to a very intense heat, much above their centres. They were fed together, the feed-pipe connecting the two boilers without means for separating them. The steam connection on top of the boilers was the same; i.e., there was no valve between the boilers. They were used in this condition for more than seven years without any irregularity in the working of the water that attracted attention; and then, without any change whatever in any of the arrangements, or in the water, other than what was caused by the introduction of something that had accumulated in the water-pipes, the water began to work in a very remarkable manner. When I examined the boilers, the putting on of the feed-pump, when the water in them was at or a little below the middle gauge-cock, would be followed in a little time by a rush of water from one boiler into the other, so as to be solid water at the upper gauge-cock in the latter boiler, and dry steam at the middle cock in the boiler from which it went; and then by a return, so that it would be solid water at the upper cock of the first boiler, and dry steam at the middle cock of the other; it would then settle down level in the two boilers, and show no sign of fluctuation by the variation of the intensity of the fires on the opposite sides of the furnace, such as shows itself when the steam connections of such boilers set together are too small.

The first time I saw this operation of the water, it went, soon after the starting of the pump, first from the left-hand boiler into the right, then back into the left, and then settled down quietly. The second time, a few hours after, the starting of the pump was followed by a rush of water from the right-hand boiler to the left, and then back to the right, and then finding its level, showing that the action was not caused by any peculiarity in the feed-pipe, which diverted the water into one boiler. This was also shown by the fact that the starting of the pump was not followed by any such action, when the water was much above the middle gauge-cocks at the time of starting it.

The boilers were tested, and found to be strong and tight; and examined, and found to be not very dirty, — that is to say, not having so much sediment in them as many boilers have without any trouble, — and no hard scale. The brick-work was found to be as indicated above, and was put down so as to confine the heat to the parts of the boilers below the lower gauge-cocks; the boilers were cleaned out, but no changes made in the pipes; and there was no more trouble with the water.

I am of the opinion that this phenomenon was produced in this way. The new element in the feed-water caused the coating of the boilers, in a way calculated to favor repulsion, but not to the extent of leading to repulsion so long as the water was up. Then as the water went down in the boilers, the one that had the strongest fire under it would have its upper surfaces uncovered faster than the other one, on account of the more rapid vaporization of its water, and because this would keep the pressure in it above that in the other boiler, thus depressing the water.* This uncovered surface would also be exposed to a greater

---

* The steam connecting-pipes were so large that this difference in pressure and water-level was very slight, except when the water was thrown over the overheated surfaces.

heat, so that the elevation of temperature would be so much greater than that of the uncovered surface in the other boiler, that the character of the coating near the front end of the boiler would be so changed by the elevation of temperature, that, when the water came up over it soon after the starting of the pump, it would lead to such a reduction of the iron to its temperature of maximum vaporization,* as to cause the throwing of the water over the overheated surface, and among the superheated steam, which would lead to a sudden elevation of temperature, and to the oscillation of the water to the back end of the boiler, so that in a very short time all the overheated surfaces in this boiler would be reduced : this rise of pressure stopped the entrance of the feed-water into this boiler, and caused the water to rush through the feed-pipe with the water going into the other boiler. The boilers were fed with water at a low temperature, and near the bottom ; and the turning of it all into the other boiler would still further check its production of steam, and at the same time reduce the temperature of the gases around its uncovered surfaces ; so that when the water came up over them, the character of the coating (in relation to repulsion) on the parts of them last uncovered had not been so changed as to lead to so sudden a reduction of temperature as to cause the throwing of the water, till the surface high-

---

* This action, taking place just at the surface of the water, would not be attended with any great local pressure, but would be sufficient to completely disperse the water over the overheated surface, and to cause its oscillation.

est up and that had been longest uncovered was reached, where the elevation of temperature had been sufficient to cause such a change in the character of the coating, as to lead to a repetition of the operation which had taken place in the other boiler, with the same results, except that the water came up over the uncovered surfaces of the other boiler before there was sufficient elevation of temperature to cause a repetition of the operation, and the water settled down to its level.

In another case, that of a boiler in which the circulation of the water was ordinarily good, the accidental introduction of a very little foreign matter, from the washing of some oily aprons, caused the water to work in a very peculiar manner. The repulsion and agitation appeared to be confined to the lower surfaces, there being no undue foaming at the surface; so that the appearance of the boiler, when I saw it at work, was such that I supposed the fact of so much water being thrown out mechanically mixed with the steam was due to the working of the water too high; but working with less water did not mend the matter, and a more careful investigation led to the discovery of the cause of the trouble.

In another instance, two sets of boilers had worked side by side for over twelve years, the feed-water being taken to them through the same pipe. The character of the water was such that there was ordinarily no trouble with the circulation when they were cleaned out twice a year. One set of these boilers was found to be leaking just after the other set had been cleaned out, and found to have very much less sediment in them than usual; in fact, they had so little in them that the other set were not cleaned out, as they otherwise would have been. After the boilers began to leak I tested them, and found they had been so overheated that nearly every seam on the bottoms leaked. They were then examined inside, and the amount of sediment found to be larger and of a different character than was ever found in them before; showing clearly, as there had been no change whatever in the pipes, that a deposit of sediment somewhere in the water-pipes had been started when this set was being fed, and had all gone through into it. It did not lead to so marked a change in the working of the water on the surface as to attract attention, and its presence was unsuspected.

Two plain cylinder boilers, sixteen feet long, and three feet in diameter, set side by side, with one furnace under them in which was a very strong fire, were found to work very badly when first started up: these boilers were supported by a cast-iron front, between which and the fire was a lining of fire-bricks. The back ends were supported by a wall about twelve inches thick, and the sides were exposed to the direct action of the flame to a point about four inches above the centre, a little below the lower gauge-cocks. The feed-water was introduced near the back ends, and was always at the temperature due the pressure on the boilers. After a few days' running, during which the water continued to work badly, the boilers were stopped and thoroughly cleaned out; and I then fired them as hard as they could be driven without a sign of unsteadiness in the circulation of the water for some hours after the steam was got up; but then, as the brick-work got heated up, the circulation became more and more unsteady; and, in about fourteen hours after the starting of the fires, the brick-work had become so heated, that the heat imparted by it to the boiler caused such an interference with the descending currents, that the water would be thrown off by a fire of not over half the intensity that could be worked, without a sign of starting, when the brick-work was cool. The thickness of the back supporting-wall was increased, so as to offer more protection to the descending current there, and the brick-work on the sides brought down, so as to keep the heat from the descending currents there, till they could acquire sufficient momentum to carry them well down; the making of these improvements in the brick-work protected the descending currents, so that the water worked very well.

Two tubular fire-box boilers, in which the provision for the circulation of the water was very good, that were worked with strong fires, had the circulation of the water so injured by the interference with their descending currents, caused by the partial filling up of their back water-legs by a loose scale from their tubes and waists, that a very large amount of water was thrown out mechanically mixed with the steam, but the water went through the engine without noise: *

---

* It was a Corliss engine, and the large exhaust-ports on the bottom of the cylinder doubtless favored this.

the increase in the consumption of fuel was not noticed, because of changes in the machinery driven ; and there were no such indications on the surface of the water, of the trouble within the boiler, as to attract attention, so that their dangerous condition was not suspected.

While the boilers were in this condition, it became necessary to learn the power required by the machines in the several departments ; and indicator-cards, taken for this purpose, showed that, when the machinery was all on, so as to require the engine to take steam very nearly half-stroke, if from any cause the pressure on the boilers was being reduced, the amount of water thrown over mechanically mixed with the steam was so large, that steam given off by it, because of the reduction of pressure within the cylinder after the closing of the steam-valve, was sufficient to raise the pressure at the end of the stroke nearly sixteen per cent. That this great overpressure at the end of the stroke was not an appearance only, caused by a defect in the indicator, or that it was caused by a leak in the steam-valves greater than that of the piston and exhaust-valves combined, is shown by these facts : that the indicator used was a very good instrument, and perfectly free at the time ; that the curves on the cards are not such as would be made by a leak of the steam-valves, — i.e., they run down regularly till the pressure is reduced so as to cause the water in the steam to begin to give off steam, and show the irregular line, while this was being done, to near the end of the stroke, where they again become regular ; and that a reduction of the load, letting the engine cut off quicker,

— so that the interference with the descending currents, caused by the reduction of pressure when the engine took steam, was less, — was followed by a reduction of the overpressure at the end of the stroke: the throwing off of about twenty per cent of the load reduced the overpressure at the end of the stroke to a little less than seven per cent. And the engine, which was only fourteen-inch cylinder, four-foot stroke, making fifty-two revolutions per minute, with ninety-seven pounds pressure on the boilers, was driving, when the machinery was all on, over one hundred and thirty indicated horse-power, and an average of nearly one hundred and twenty indicated horse-power through the day, with the consumption of but little over three pounds of ordinary anthracite coal per indicated horse-power per hour; and this with a safety-valve on the exhaust-pipe weighted so as to cause a pressure of over one and one-half pounds per square inch (the pressure required to do the warming and drying) in the exhaust-pipe.

We now come to the consideration of the class of explosions in which there is the same shattering of the upper part of the shell as in the class last considered: where there is no evidence of overheating, the explosion occurring after the boilers have been quiet for a greater or less time, with very moderate fires in them; and, in some instances, when it is known that the pressure, but a moment before the explosion, was very low, — i.e., with the fires so moderate as to preclude the idea of repulsion, and with too low pressure of steam and consequent temperature of water (unless the water was at a temperature above that due the pressure) to shatter the boiler, even if there was a large break in the top.   In one instance of this kind, — of a locomotive less than one year old, — the pressure, as shown by the steam-gauge, was noticed by the watchman but a moment before the explosion to be but forty pounds.   In this case, the engine had been run into the house with the water well up, the night before, and had so stood through the night till the time for starting the fire for the morning train.   The fire had been started, and the engine was standing with this moderate fire and low pressure of steam, when the explosion occurred.   The fireman, who was standing by its side, was killed.   The watchman, who had just been in conversation with the fireman, and had then observed the pressure, had started away, and escaped injury.

This was an explosion of terrific violence: the top of the shell was torn completely open in all directions from a point nearly over the fire-box.   The most careful examination did not show any defect in construction or material, that would account for the yielding of the upper part of the shell at a pressure less than one vastly above that on the boiler a moment before the explosion.

In another instance a boiler had stood with the fires banked up over night.   The fireman, who was killed by the explosion, had orders to get up steam early in the morning: the time of his going to the fire-room was known; and in a time after this so short as to preclude the possibility of any great rise of pressure, the boiler exploded with great violence.

Mr. Colburn reports one case of this class: " On the 12th of February, 1856, the locomotive ' Wauregan ' exploded, after standing for upwards of two hours in the engine house of the Hartford, Providence, and Fishkill Railroad, at Providence, U.S. Only sufficient steam had been maintained in the boiler to enable the engine to be run out of the house ; but at the time of the explosion the engine had not been started, the engine-man, who was killed, being upon the floor, at the side of the engine, at the time. The boiler gave out in the ring of plates next behind the smoke-box." Mr. Colburn in this connection adds, that " destructive explosions often occur at pressures of ten pounds to twelve pounds per square inch in low-pressure boilers."

In relation to this class of explosions, it is clear that a boiler may be so weak as to yield to a very low pressure, or that a strong boiler may be broken when the pressure is very low, by water thrown after repulsion ; but it is also clear that the steam given off by water, on a reduction of pressure above it, cannot have a pressure above that due the temperature of the water ; so that in case of the rupture, from any cause, of the shell of a boiler in which the pressure is low, and in which the water has the temperature due its pressure, it does not appear possible that the water and steam will be thrown with such force as to shatter the shell, even if the rupture is very large ; in fact, unless the opening is very large, the escape of steam will be so slow that the reduction of pressure will be too gradual to lead to any such destruction as is frequently caused by these explosions. And I am of the opinion that all violent explosions occurring when the fire is moderate and the steam low, are caused by the explosive giving-off of steam by water that has been heated up to a temperature above that due the pressure. M. Magnus and M. Donny have each demonstrated that water may be heated in an open vessel to a temperature much above its ordinary boiling-point.

Mr. Bourne, in the eighth edition of his Treatise on the Steam-Engine, says (pages 79 and 80) in relation to these experiments: "M. Magnus found that water well cleared of air may be raised to a temperature of 105° or 106° C. (222° F.) before boiling. . . . M. Donny, however, by freeing the water more carefully from air, succeeded in raising it to a temperature of 135° C. (275° F.), without boiling; but at this temperature steam was suddenly formed, and a portion of the water was suddenly projected from the tube. M. Donny concludes, from his experiments, that the natural force of cohesion of the particles of water is equal to a pressure of about three atmospheres, and to this strong cohesive force he attributes the irregular jumping motion observed in ebullition, and also some of those explosions of steam-boilers which heretofore have perplexed engineers. It is well known that cases have occurred in which an open pan of boiling water has exploded with fatal results, and such explosions cannot be explained on the usual hypothesis. M. Donny says that liquids by boiling lose the greater part of the air which they hold in solution, and therefore the molecular attraction begins to manifest itself in a sensible manner. The liquid consequently attains a temperature considerably above its normal boiling point, which determines the appearance of new air bubbles, when the liquid separates abruptly, a quantity of vapor forms, and the equilibrium is for the moment restored. The phenomenon then recurs again with increased violence, and an explosion may eventually ensue."

Professor Miller, of Kings College, London (see his "Elements of Chemistry," Part I. page 224), says, in relation to this matter: "By long boiling of the water the air becomes nearly all expelled: in such a case the temperature of the water has been observed to rise as high as 360° F. in an open glass vessel, which was then shattered with a loud report, by a sudden explosive burst of vapor. In this case the force of cohesion retains the particles of the liquid throughout the mass in contact with each other, in a species of tottering equilibrium; and when this equilibrium is overturned at any one point, the repulsive power of the excess of heat stored up in the mass suddenly exerts itself, and the explosion is the result of the instantaneous dispersion of the liquid."

In one instance in my experience, in which I am confident now (although at the time it was to me, and to other engineers with whom I talked, a perfect mystery,) that there was such an elevation of the temperature of the water, the facts are as follows: There were eleven boilers in all in the house, plain cylinder boilers, thirty inches in diameter, and a little over forty feet long. One set, of four of these boilers, were side by side, with one furnace under them. They were connected below by the feed-pipe, and fed together, and connected above, at a point rather back of their centres, by means of a drum of about the same diameter as the boilers, and of a length sufficient to bring its ends over the side walls of the setting, so as to serve as a means of support to the centres of the boilers. Each boiler had an eight-inch nozzle riveted on the top, and the drum had four nozzles of the same size. These nozzles had flanges, by means of which they were bolted together. There was also a connection by means of four-inch nozzles and bent copper pipes, just over the furnace at the front ends of the boilers, for two safety-valves. The expansion and contraction of the drum, together with the weight of the boilers, caused the nozzles on the drum and on the boilers to leak. There were boiler-plate gaskets between the nozzle flanges and the boilers, and between the nozzle flanges and the drum, and the rivets were "driven" on the outside; the leaks were around these gaskets, and they required frequent calking. One Saturday night, after the fires had been banked up, I closed the valve between the drum on this set of boilers and the main steam-pipe, took in a good quantity of water, blew off steam, and calked these gaskets, leaving the boilers in this condition. The heat of the banked-up fire and of the brick-work was such that the steam would probably continue to rise slowly for ten or twelve hours, when it would be at or near the blowing-off point, — eighty pounds per square inch; and then after standing for some time nearly stationary, the pressure would slowly fall, so that by the time for starting the fires Monday morning it would be about fifteen to thirty pounds. The

boilers stood in this condition till about four o'clock
Monday morning, perfectly quiet, when the fires were
drawn forward and cleaned out, but, before they had
got to burning up at all strong, and when it was supposed the pressure was about the same as that in the
main pipe and the other boilers, and probably not over
forty to fifty pounds per square inch, the starting open
of the valve at the drum was instantly followed by a
very severe shock at the front ends of the boilers, and
the bursting of one of the four-inch copper pipes connecting the boilers at the front ends, from which water
was thrown. The water then settled down, and all
was quiet. The pipe burst showed no sign of imperfection in the brazing, and I suppose it would have
stood a steady pressure of three hundred pounds to
four hundred pounds per square inch.

These boilers were nearly new, and of very tough
one-fourth inch iron, and, taking into account their
diameter, they were very strong at the front ends, so
that, although the blow from the water was of tremendous force, the shells were not broken. The blow appeared
to be confined to the front ends ; there was no appearance that the water struck up into the drum or against
the shells around the big nozzles, which, because of
the reduction of strength there, caused by the cutting
out for the nozzles, and the previous strains and leaks
would have been shown by the opening up of the old
leaks, but was not the case ; and I am now of the
opinion that the water just over the fires was heated
up to a temperature very much above that due the
pressure, and that the agitation produced by the start-

ing open of the valve caused it all to give off its steam, throwing the water above it with great force against the shells of the boilers over the furnace, and that this overheating was confined to the water over and very near the fires.

The water (from Merrimac River) used in these boilers was heated by being showered into the top of a vertical tank, into which the engines exhausted, so that the tallow used for the lubrication of the valves and pistons of the engines went into the boilers, and it was supposed that the presence of this grease might possibly have been the cause of the explosion. But an examination, after the explosion, showed the boilers to be clean, with the exception of a little sediment at the back ends, which showed the presence of the tallow, and a very little greasy coating near the water-level. These boilers had to be driven hard to make steam for the engines, with a strong draft; but there was never any trouble in the working of the water when the fires were strong. That overheating of the surfaces of the boilers above the water, or superheated steam, was not the cause of the shock, was shown by the fact that the water was more than four inches above the highest point exposed to the direct action of the heat. This was in 1847.

In 1850, I had the charge of two sets of boilers, exactly alike; each set was composed of five boilers, thirty-three inches in diameter, and forty-one feet long, fed together, a two and one-half inch feed-pipe connecting them at the front end; the steam connection was by means of a drum, of the same diameter as the boilers, running across the five, with eight-inch nozzles on the boilers and drum. The water was got up in each set of these boilers one Saturday night, so as to show at the upper gauge-cock; and soon after this the fires were banked up, and the steam-valve between one set of boilers and the main pipe was closed. From the other set, steam was being drawn to two quite large dry-rooms. After the boilers had been standing in this way for several hours, the pressure on the set from which the steam was being drawn, probably very nearly as fast as it was made, was found to be nineteen pounds per square inch, and the pressure of the other set to be twenty-nine and one-half pounds per square inch, by the same gauge, a U mercury gauge. The valve connecting the set of boilers in which the pressure was twenty-nine and one-half pounds was then opened very slowly, but the rise of pressure on the main pipe was so sudden as to cause the mercury, standing at nineteen pounds pressure, to be thrown to very near the top of the tube of the gauge used, which was a sixty-pound gauge, with more than six inches spare length of tube on the upper end. The mercury then fell, and some of it went over into the drain-pipe, but just how much could not be known. After a few oscillations, the mercury came to rest, and stood at twenty-nine pounds strong, — that is, within less than one-half pound of the pressure in the boilers where the steam was highest. To do this, the steam given off by the water in these boilers had to raise the pressure within the other set of boilers, and in the drum on them and in the main pipe, which, with its branches, amounted to nearly one hundred feet of eight-inch pipe, one hundred and fifty feet of six inch, forty feet of four inch, forty feet of three inch, sixty feet of two inch, and over two thousand feet of one inch pipe in the two dry-rooms; and also to raise the temperature of a thin film of water (over five hundred square feet) in the set where the pressure was lowest; all the iron above the water in this set, its drum and nozzles, and the main pipe, over 17° F., showing that a very large volume of steam must have been given off by this water.

The water (from Nashua River) used in these boilers contained a vegetable matter, and an earthy matter, that coated the interior surfaces in such a manner, that, when it was let out of them, when the brick-work was hot, the coating hardened into a scale; but if the brick-work was cool when the water was let out, it was found in the form of a brown, slimy coating over the surfaces.

The method of heating the feed-water was such that no grease from the engine went into the boilers, and there was never any trouble with the working of the water in them when the fires were driven hard, with a very strong draft. There was no superheating of the steam, or overheating of the iron of the boilers above the water.

In another instance, two tubular boilers, which were exactly alike in all respects, set at the same time side by side, but having separate furnaces and connections, so that one could be used without the other, but which had always been run together, were fed one at a time with water that was heated in the same surface condenser. These boilers were run under a pressure of ninety-eight pounds per square inch, making steam for an engine that was so heavily loaded, that a very slight fall in pressure would cause it to take steam full stroke, but the circulation of the water was so good in them that there was never a sign of trouble with it, even when subjected to this trial, with the water showing at the top gauge cocks. When these boilers had been in use about three years, one of the check valves was found to be stuck up with a sort of slimy coating, so that in getting it out the spindle was broken. The valve was repaired, and nothing peculiar was noticed in the working of the boiler when the fires were hot, but after they were banked up nights, and when the steam, which was used from the two boilers for warming the building nights, had been run down to about forty pounds per square inch, and when the fires were so dead on top that the pressure — as shown by a gauge connected with the two boilers — when no steam was being drawn from them, rose only at the rate of from one eighth to one fourth of a pound per minute, this boiler gave unmistakable evidence that the water was being heated up to a temperature above that due the pressure, and that its temperature was then reduced by an explosive generation of steam. The shocks produced by these explosions were reported to me by the engineer, to be so severe as to shake the boiler and its setting. The boiler was very strong and no leak was started in it. There were no indications of this trouble in the other boiler, which was being subjected to exactly the same treatment, so far as could be seen. The boilers were cleaned out, and very little sediment found in them, with no noticeable difference in the character of that in the two boilers. But still, after the boilers had been standing as above, there were the old indications of trouble, only not so heavy ; and a second thorough washing of the boiler entirely removed the cause of the trouble.

The method of heating the feed-water in this case, was such that no grease from the engine could get into it, and there was no superheating of the steam. What the effect of opening of a valve so large as to have caused much of a reduction of the pressure above the water would have been, was not known, the branch leading to the warming pipes being so small that the opening of its valve would not cause anything like a sudden reduction of pressure.

In each of these three cases, it will be noticed that there was no trouble with the water when the boilers were being driven hard, and in neither of the two first was there the least indication of trouble, such as the shocks in the last one, till the reduction of pressure caused by the opening of the large valves.

Very little is apparently known as to the manner in which the water becomes so freed from air, or how the little particles of solid matter in it are so precipitated as to lead to such a heating up. It will be noticed, that M. Magnus appears to have been unable to free the water used by him of air, so as to raise it to a temperature in the open air above about 222° F., while M. Donny succeeded in reaching a temperature of 275° F. Professor Miller gives an instance of a temperature of 360° F. having been reached; and M. Donny gives cases of explosions of open pans of boiling water, " with fatal results." From which it appears to me to be probable that the freeing of water from air depends very much upon the character of the water used, so that the treatment that would expel but little of the air from some waters, would cause it to become so far expelled from others, as to lead to serious results. So far as I am aware, very little is known of the effect in this respect, of boiling under pressure. It is known that some waters may be boiled in a vacuum for some time without completely expelling the air; but, so far as I am aware, it is not known that the same water might not have its air expelled in a steam boiler working under a strong fire and high pressure, the elevation of temperature of the water, doing with the increase of pressure, what could not be done by the removal of the pressure of the atmosphere, at the consequent low temperature; or, that water that has been boiled under a high pressure with its consequent elevation of temperature, may not have its air so far expelled that it may be in condition to be raised without boiling by a slow fire, under a pressure very much lower, to a temperature nearly as great as that to which it had before been raised. Another consideration, and one which I think may have much to do with the matter is this, — that water in a steam-boiler may be so far freed from air, as to be in a condition to have its temperature raised very much above that due the pressure,

but for the presence of little particles of solid matter
in it; and that when the conditions of the water are
such, that these little atoms of solid matter are all
precipitated out of the immediate reach of the fire by
the action of a moderate heat, the continuation of the
same heat may then raise the temperature of the water,
either till it gives off steam explosively because its
molecular attraction is overcome, or till some agitation
of the water, or an increase of the heat, may throw up
the little particles of solid matter so that they shall
become mixed with the overheated water, and so cause
the excess of heat to be given off in an instant.

However this may be, or whatever may be the causes
of the heating up of water in steam-boilers to a temperature above that due the pressure, I think there
can be no doubt of the fact, and that it is the cause
of all the explosions in which the shells of boilers
are shattered, when the fires and pressure are low;
and that in such explosions the water is thrown
much in the way that it is after repulsion, but accompanied by a greater rise in pressure; and that so much
water may be overheated, and to such a degree, that
so much may be thrown, and with such force, and
accompanied by such an elevation of pressure, as to
cause the complete shattering of the shell by the first
blow.

That the water is sometimes raised to a very high
temperature in such explosions, is shown by the fact
that in several instances so much of it has flashed
into steam as to leave no sign of water about the exploded boiler. Mr. Colburn gives two instances of

this kind, and says in relation to them, — " It has, indeed, been assumed, that in many cases of explosion all the water previously contained in the boiler is converted into steam. Mr. Edward Woods once mentioned, at the Institution of Civil Engineers, an instance which came under his observation, in 1855, I believe, and where, after a locomotive boiler had burst, the whole of the water was found to have completely disappeared. Mr. Vaughan Pendred, of Dublin, has informed me that he observed a similar result after he had exploded a small boiler, well supplied with water, for the purpose of experiment. He had erected a fence of boards about the place where the boiler was allowed to burst, but on going to the spot immediately afterwards no traces of water could be seen."

Having now considered how boiler explosions are caused, we will turn our attention to the matter of how they may be prevented; and, in the consideration of the matter of preventing boiler explosions, it is to be remembered that the aggregate loss, aside from the loss of life, from the use of defective boilers that do not explode, is very much greater than the direct loss from explosions. The number of boilers that explode is but a small percentage of the number built; but the number of boilers in use, in conditions that cannot be considered safe, and in which the loss — from the gradual destruction of the boilers themselves, from the waste of fuel, the injury to engines from the impurities carried over by the water mechanically mixed with the steam, and the loss from the same cause in many dyeing and bleaching operations — is great, is a very large percentage of the whole number in use. So that the saving in dollars and cents resulting from the use of means to prevent boiler explosions will be vastly greater than the value of the property destroyed by explosions.

To prevent explosions from low water, boilers should be so constructed and set, and have their gauge-cocks so placed, that no part of their shells can be over-heated when the water is at the lower gauge-cock; that there shall be sufficient steam room and area of water level to avoid priming, when the water just shows with the steam at the upper gauge-cock; and that the best steaming point shall be with solid water at the middle gauge-cock, and dry steam at the upper one. The engineer or fireman in charge, should then be held responsible for the keeping of the water at this point. There should be no dividing the responsibility between him and any automatic feeding apparatus whatever. The gauge-cocks should be so constructed that they can be cleaned out under steam, and always kept clean, so that when opened there shall be a clean passage through. If the gauge-cocks are connected with the boiler by pipes, the boiler end of the pipe must be enough lower than the other that water will not re-main in the pipe when it is below it in the boiler. If glass gauges are used, the above points should be in-dicated on a scale near the tube, as to height of water; and the connections should be so large and direct that the rise and fall of the water in the tube shall be so great that the least contraction of the area of the lower connection by sediment shall be noticed at once; and the connections must be kept clean. I do not advise reliance upon glass gauges, because of the danger connected with the filling up of the connections, the coating and loss of transparency of the tube, and because of the variation of the level of the water in the tubes from causes not easy to understand. I was once called upon to examine a boiler with which there had been a great deal of trouble, which had been attributed to priming, where this variation was so great as to render the gauge utterly unreliable. The water, when the fire was very strong, at times stood six inches above its level in the boiler. The guage was on the smoke-box cover of a cylindrical tubular boiler with the fire under its shell, and the upper connection ran through the smoke-box to the tube-plate, about on a level with the top of the glass tube; the lower connection ran through the smoke-box near the shell of the boiler to near the tube-plate, and then turned down, following the curvature of the shell to near the bottom of the tube-plate, where it entered it. When these connections were clean, so that water would be blown freely through the lower one, and steam as freely through the upper one, and both connections were opened, — when the fire was strong enough to keep up steam for the engine, the water would come up to the level of that in the boiler, and stand for a moment, and then rise suddenly in the tube about an inch, where it would stand for a short time, when the operation would be repeated, till, when the fire was strong, it would rise as high as before indicated. Whether this action was alone due to a little of an explosive boiling on the bottom of the boiler, to the elevation of temperature within the smoke-box, or both, I do not know ; but the changing of the lower connection, from near the lower part of the tube-plate to just above the top row of. tubes, cured it, so that when the connections were clean, the variation of water was not noticeable.   I have never since been so well assured that I knew where the water in a steam-boiler was, by looking at a glass gauge — no matter how well the water might appear to be working in the tube — but what I wanted also to try the gauge-cocks.

If from any cause the water gets· below the lower gauge-cock, the fire should be drawn with the ash-pit doors closed and the draft all on, so that the rush of cold air over the fire may help to counteract its effect upon the boiler ; at the same time if there is an engine running, the steam should be gradually shut off from it, while some valve that will give a continuous flow of steam from the boiler is very gradually opened, and this valve must be so regulated that the steam shall escape from the boiler just as fast, but no faster, than it is made, closing it up fully, when the temperature of the brick-work is so low that the pressure does not rise. The boiler should then stand without the doing of any thing to cause the least agitation of the water, and allowed to cool down till it can be examined and tested before it is used.   In relation to this matter the Committee of the Franklin Institute say : " If the engine is at rest, in such a case, it should not be put in motion. If it is in motion it should be slackened or stopped, the fire-doors opened, and the heat got down. The opening of the safety-valve in such a case should be avoided. The engineer should remember that as life is at stake he cannot be too prudent." To call attention to the fact of low water, and to help reduce the intensity of the heat in such a case, I advise the employment of a suitable low-water safety-plug in the crown-sheet, or (if a cylinder boiler with the fire under it) in the shell over the fire, and just below where the brick-work comes over against the boiler. The reduction of temperature, by such an escape of steam into a fire, is of very great value as a preventive to the burning of the boiler in case of low water. In one instance, a large fire-box boiler was saved by its safety-plug when its blow-off valve was accidentally opened wide, when there was a heavy anthracite coal fire on. The reduction of temperature, by the blowing of the steam into its furnace, was so great that the fires were got out without the starting of a leak in one of its two hundred and thirty two-inch tubes.

There is great misapprehension in the minds of many persons regarding the danger of overheating the upper part of the shells of boilers, from the fact that in many instances boilers have had the upper parts of their shells exposed to the action of the gases, after they had passed through the flues or tubes, without serious results. In many instances tubular boilers, in which the circulation was so poor that much water was thrown up mechanically mixed with the steam, have had the consumption of fuel reduced by taking the gases, after they had passed through the tubes, over the tops of their shells, retaining sufficient heat to serve to vaporize the water so thrown up, with the con-

sequent reduction of fuel, but not sufficient heat to
cause any great heating up of the shell of the boiler,
or superheating of the steam, even when the shells
were clean ; and in many instances of this kind I have
examined, the parts of the shells so exposed have
been found to be so covered with fine ashes as to be
pretty well protected from the action of the heat ; and
I think many more boilers would have had their shells
broken by the heating up of their tops than have, but
for this accumulation of fine ashes. And no matter
how much water may be thrown up mechanically
mixed with the steam, attempting to vaporize it by
heating up the top of the shell is an unsafe operation.

In relation to the heating up of a boiler above the
water, the Committee of the Franklin Institute say :
" In a former division of the subject, the Committee
showed the great danger which is produced in a boiler
by highly heated metal ; any boiler, therefore, which
has parts exposed to heat without being in contact
with water, is essentially defective."

And the experiments of the Committee demonstrated
perfectly that the application of heat to the top of the
shell of a boiler, with a view to the superheating of the
steam contained in the boiler, is only attended by a
heating up of the metal of the boiler, and a very little
of the steam in its upper part. In several instances,
during these experiments, the steam was superheated
to a temperature as high as 528° and 533° in the top
of the boiler, and yet the injection of water at a temperature below 212° into the steam within the boiler,
but a few inches below the point where the thermom-

eter indicated the above temperature, did not reduce the temperature of the superheated steam, so as to cause the thermometer to fall in a single instance. So that the vaporization of water thrown up mechanically mixed with the steam, by taking the hot gases over the top of the shell — even though so much water is thrown up, that while steam is being drawn from the boiler there is no great rise in the temperature of its top — is always attended with this danger: that, whenever the demand for steam ceases, the top of the shell of the boiler, and a thin film of steam within it, will be quickly raised to an unsafe temperature. The vaporization of this water in the boiler can only be done with safety in vertical " fire-tube " boilers, in which the shells above the low-water point are not exposed to the action of heated gases, and in which the tubes are so small, and their length such, that the water can be kept so high, without unduly contracting the steam room, that there shall be just enough heat left in them to vaporize the water thrown up; that is, that the temperature of the gases within the tubes at that point, shall never be enough above that of the water within the boiler to cause a superheating of more than a very few degrees, even when the steam is shut into the boiler, when the fires are strong. In all other boilers the true remedy is to so improve them that the amount of water thrown up shall be reduced so low that the steam can be used to advantage. While on this matter, it may not be out of place to say, that the benefit to be derived from superheating, is a very questionable one. It is not safe to use superheated steam for warming buildings, because of the danger of fire. Probably a very large proportion of all fires caused by steam-pipes have had their origin in superheated steam. In one instance of this kind that I investigated, where the fire was set over forty feet from the boiler, there were unmistakable evidences that the steam was superheated. And whatever theoretical advantages appear to attach to the use of superheated steam in a steam engine, in practice, the superheating of high pressure steam does not pay. The specific heat of steam is so low, that a superheating that will cause the serious cutting of valves is so far reduced by radiation from the steam-chests and cylinder-covers, as to be of very little use in the cylinder after expansion commences; and a little condensation on the wearing surfaces of the interior of a high pressure steam-engine, appears to me to be the cheapest way to lubricate them, and, in fact, the only way in which an engine of any size can be kept from cutting.

Explosions of properly constructed boilers, by a steady increase of pressure, may be prevented by the employment of suitable safety-valves and safety-plugs. The plugs are to be so constructed, that in case of the failure of the safety-valve from any cause, they may be broken before the pressure gets so high as to endanger the boiler, and placed where they will be acted upon by the most intense heat of the fire, so that, in addition to the pressure of the steam, they shall have their strength reduced by the elevation of temperature, and be beyond the reach of accidental or wilful interference; so placed also that the leak caused by their breaking shall so far reduce the temperature of the fire, as to prevent a greater rise in pressure, even if the leak is not observed. The valves, and seats of the safety-valves, should always be of composition; the seats should either screw into the cases, or be secured by suitable pins; and the spindles, or wings of the valves, should be so much smaller than the holes in the seat, as to render it impossible that they shall become fixed by the greater expansion of the seat than of the case, as mentioned on page twenty-one.

Explosions from defective materials and construction, may be prevented by a proper testing and inspection of materials and of boilers; so that any defect in either shall be discovered by the tests, and not left to cause the explosion of the boiler. In testing the plates, great care should be taken to detect imperfect welds. Boilers in use should be examined very often, by scraping off the soot from any spot, having a different appearance from that of the surrounding surfaces, with a blunt-pointed instrument, in order that the first parting off of an imperfect weld may be detected. Such blisters, if taken in time, may be riveted down so as to make a very much better job than can possibly be made by patching. Imperfect welds of this kind usually show themselves, in new boilers, soon after heavy fires are got up; but in one instance in my experience, where the boilers were fed on the bottom over the fire, a weld that had stood for over a year, while using hot feed-water, was parted by feeding the boilers with cold water. When boilers are fed on the bottom over the fire, the feed-water should always be as hot as possible, not only on account of the saving in fuel, but to avoid the dangers connected with the sudden reduction of the temperature of the bottom of the boiler. And it should be taken in slowly. Some engineers suppose, that by an occasional rush of feed-water into the bottom of a boiler over the fire, sediment that has settled there may be washed back; but it is exceedingly doubtful whether any substance heavy enough to settle over the fire is ever started in this way; and there can be no doubt whatever in relation to the strain thrown upon the boiler by the operation. The best place to introduce feed-water is through the shell, at some distance from the fire, and in such a manner that it shall become so mixed with the descending currents as to have its temperature raised to very near that of the water within the boiler, before it comes in contact with the shell of the boiler. And it is advisable on all accounts to construct boilers of a quality of iron that will give the requisite strength, without having to resort to the use of very thick plates. Regarding the matter of testing boilers, there is a feeling on the part of some, that the testing of a boiler, by forcing water into it with a pump, injures it so that a boiler that stands a water-test of a given pressure may have its strength so reduced by the operation, as to yield immediately after to a steam pressure very much lower. It is undoubtedly true, that a boiler which stands a water-test without a sign of yielding may explode soon after at a very much lower pressure, because of repulsion, low water, or the overheating of the water; for a boiler may be so defective in relation to adequate provision for the circulation of the water, or in relation to the exposure of surface above the water-level, as to be utterly unsafe to use, and still be strong enough to stand the required water-test perfectly; and such a boiler might explode without the previous testing having any thing whatever to do with it. And while it is also undoubtedly true that many boilers have been injured in testing, by parties not qualified to perform the operation, — which is one requiring a knowledge of the strength of materials, and one that must be performed with much care, — it must be borne in mind that testing with water is the only safe way to learn that there are not hidden defects in the material, that may disclose themselves by explosion the first time steam is got up on the boiler; or that there are not such alterations of form, as will lead to the destruction of the boiler by the constant bending under varying pressures, or that may cause leaks that will lead to its destruction by corrosion, in new boilers; or that a boiler in use has not had its strength so far reduced by corrosion, by unequal expansion, or by alteration of form, as to render it unsafe; and that, when properly performed, it is not an operation attended with the least injury to the boiler.

And now we come to the question of what is the proper strength of a steam-boiler for any given pressure, and to what water-pressure it should be subjected before it is set. I am of the opinion that all boilers, intended for working under a pressure of forty pounds per square inch, or less, should have a calculated bursting strength of three hundred and fifty pounds per square inch; and that all boilers intended for a working pressure above this, should have this calculated bursting strength increased by two and one-half times the additional working pressure; so that a boiler, to be worked at a pressure of seventy pounds per square inch, should have a calculated bursting strength of four hundred and twenty-five pounds per square inch; and one to be worked at a pressure of one hundred pounds per square inch, should have a bursting strength of five hundred pounds per square inch; and that all boilers should be tested, when new, and before they leave the shop, with water at a temperature of 60° to 70°, up to a pressure two-fifths of the calculated bursting pressure: — that is, a boiler to be run under a pressure of forty pounds per square inch, to one hundred and forty pounds; one to be run under seventy pounds, to one hundred and seventy pounds; and one to be run under one hundred pounds, to a pressure of two hundred pounds per square inch; and that, at these test pressures, there should be no perceptible alteration of form, or leak that will not rust up perfectly tight before the boiler is used, without the putting of any thing whatever into it to cause rusting, or filling up of defects in workmanship. The boiler should be properly supported before it is filled with water, and, as it is filled, the air should all be allowed to escape from the highest point of the boiler, so that it may be full of water. The effect of every stroke of the pump upon the gauge should be observed, and there should be no difference in the temperature of the water in the top and bottom of the boiler. Boilers, after being put to work, should be tested once a year, and oftener, if it is known that they have been subjected to an undue strain, or overheating, at a pressure three-fourths of that indicated for new boilers. That is, an old boiler, working under forty pounds pressure, should be tested to one hundred and five pounds; one under seventy pounds pressure, should be tested at one hundred and twenty-eight pounds; and one working under one hundred pounds, at one hundred and fifty pounds: and this should not be a mere putting on of the pressure, and a passing of the boiler because nothing breaks, but a thorough examination for alteration of form, or other indications of weakness; and if any such are found, the boiler should either be strengthened, or the working pressure should be reduced; and if leaks are found, they should be stopped by calking or patching; never by the putting of substances into the boiler to fill them up. There are cases where a boiler is known to be weak, when a more frequent testing at a lower pressure may be advisable, till a change can be made without serious loss. In testing old boilers, particular care should be taken as to the temperature, that their tops are not above that of their bottoms, when tested; they should always be cleaned out after being tested. In relation to the testing of boilers, the Committee of the Franklin Institute, say, in a form of law relating to the matter, submitted with their report, " And he (the Inspector,) shall, moreover, provide himself with a suitable hydraulic pump, and after examining into the state and condition of the boiler, or boilers, it shall be his duty to test the strength and soundness of said boiler, or boilers, by applying to the same a hydraulic pressure equal to three times the certified pressure which the boilers are to carry in steam." Professor Rankine says, in relation to the strength of boilers, and the testing of them : "Before any boiler is used, its strength ought to be tested by means of the pressure of water forced in by pumps. The *testing pressure* should be *not less than double the working pressure*, and *not more than one-half the bursting pressure ;* that is to say, as the bursting pressure should be six times the working pressure, the testing pressure should be between twice and three times the working pressure. About *two and a half* times the working pressure is a good medium. A pressure of water is to be used in testing boilers, because of the absence of danger in the event of the boiler giving way to it."

In relation to the great importance of attention to the matter of the strength of boilers, Mr. Fairbairn says : "It appears to me equally important that we should have the same proofs and acknowledged system of operations in the construction of boilers, that we have in the strength and proportions of ordnance. In both cases we have to deal with a powerful and dangerous element; and I have yet to learn why the same security should not be given to the general public as we find so liberally extended to an important branch of the public service. In the ordnance department at Woolwich, (with which I have been more or less connected for some years), the utmost care and precision are observed in the manufacture of guns; and the proofs are so carefully made under the superintendence of competent officers, as to render every gun perfectly safe to the extent of one thousand to one thousand two hundred rounds of shot.

"Boilers and artillery are equally exposed to fracture, and it appears to me of little moment whether the one is burst by the discharge of gunpowder or the other by the elastic force of steam. Surely boilers are equally if not more important, as the sacrifice of human life appears to me to be much greater in the one case than in the other. It would be a matter of paramount importance to the public, if men, combining the greatest practical skill with the highest scientific attainments, would give such an *undeniable security* to boilers, as to insure them capable of bearing, under the most unfavorable contingencies, at least *six times* their working pressure."

To prevent explosions caused by scale and sediment, steam-boilers should be so constructed, that every part of their interior surfaces can be reached for their removal. And they should not be suffered to accumulate so as to exclude the water from the iron; and to guard against overheating caused by the sudden and unsuspected accumulation of heavy scale, safety-plugs should be put in the parts of the boiler exposed to the most intense heat, of such a character that they shall yield, before the elevation of the temperature of the boiler is such as to be dangerous.

The formation of scale and the accumulation of

sediment may be reduced by a judicious system of blowing off from parts of the boiler where the water is quiet, and where the sediment that settles in still water may be reached; and a good deal of the surface scum may be got rid of by the use of pretty large gauge-cocks, and, every time the cock at or just below the surface is opened, letting a quantity escape according to the condition of the surface. The blowing off from the bottom should be when the boiler is at work, and never so much as to expose any part of the surface to a temperature above that due the pressure. The water should never be let out of a boiler, set in brick-work, while there is heat enough in the brick-work to harden the slime left on the surfaces into scale. In many instances where very hard scale has been formed, while the practice of blowing the water out under steam was followed, the cooling down of the boiler before letting the water out, caused a complete change, a slime that could be washed off with the hose, being found, instead of a hard scale. Aside from all consideration of the danger of explosion, the matter of so constructing boilers that every part of their interior surfaces can be reached for the removal of scale and sediment is of the *utmost importance*. For notwithstanding the fact that these coatings are in many instances of such a character, that they are not removed from the surfaces even by a great elevation of temperature so as to cause explosions, and also of such a character that the water is either not repelled from them, or if repelled, their rate of maximum vaporization is so low that explosion does not follow its

return to them, — still the elevation of temperature of
the boiler, because of the non-conducting coating, is
attended by a more or less rapid destruction of the
boiler, and with a great waste of heat, and of loss of
power of the boiler. And when the surfaces are so
covered, it can never be known that so much of it may
not part off as to lead to the breaking of the boiler, if
not to an explosion.

Explosions caused by repulsion may be prevented
by so constructing boilers, that the circulation of the
water in them shall be so good, that it shall be brought
into forcible contact with every part of their surfaces
exposed to great heat, and to this end they must be so
constructed that every part of their surfaces can be
reached for the removal of any sediment or coating that
interferes with the circulation of the water, so that they
may be kept clean ; and by the employment of suitable
safety-plugs in the parts of the boiler from which the
water is most likely to be repelled, so that in case of
repulsion the breaking of the plug may cause so gradual an escape of the steam as to prevent the too sudden
cooling off of the surfaces, and so save the boiler.
In case of indications of a repulsive action, nothing
should be done to cause a sudden rise of pressure,
or the sudden cooling down of the parts of the boiler
exposed to the fire ; but the fire should be checked by
the closing of the damper, and the steam should be
allowed to escape from the boiler by a steady flow, a
little faster than made, till the fire and boiler are cooled
down and the boiler examined. In such a case, the
opening a little of the blow-off cock at the front end

of the boiler will favor the gradual reduction of the temperature of the boiler, and so will be an advantage ; but this must be done with *great care*, and but *very little*, or it will do more harm than good. No feed whatever should be taken in till the boiler is cooled down.

And here again we find that the construction of boilers, so that the circulation of water within them is good, is of very great importance aside from all consideration of the danger of explosion. For poor circulation of the water is always attended with a more or less rapid destruction of the boiler, and in many instances this destruction of the parts of the boiler exposed to the most intense heat has been very rapid, and yet without explosion, because of the presence of a scale or coating to which attention has already been called, which so far reduces the vaporizing power of the surfaces that the return of the water to them is not followed by an explosive generation of steam. Mr. C. Wye Williams, — in his " Elementary Treatise on Combustion of Coal, and the prevention of Smoke : " John Weale, London, 1858, — gives the case of the " Great Liverpool," on her first trip to this country in 1842 : " The engineer observing the side plates of the furnaces continually giving way, some bulging, and others cracked and leaking, and some even burnt into holes, — although there was always a sufficient height of water in the boiler shown by the gauges — supposed there was something interfering to keep the water from the plates ; and with a view of testing it introduced an inch iron pipe from the front into the water space between two of the furnaces. This at once brought the evil to notice; for, although the glass gauge always indicated a sufficient height of water, yet nothing issued from the pipe but steam as long as the boiler was in full action. The overheating was the result of insufficient circulation depriving the deep narrow flue spaces of an adequate supply of water." When this boiler was examined inside, the passages for the descending currents were found to be partially filled with scale and sediment.

Poor circulation of water within a boiler is also always attended with the throwing up of water mechanically mixed with the steam. Probably there are no boilers in use but what throw up water in this way; but in some this is so small as to be of no practical importance. Yet in many instances the quantity of water thrown up is so large as to be (as has before been mentioned) attended with serious loss; and to attempt to remedy such defective boilers, by bringing the gases up on their shells, is an operation attended with too much danger to be advisable. No matter how good the combustion may be, or with however little excess of air, or however low the temperature of the gases may be when they leave the boiler, good results can never be got when the circulation of the water is so defective that much of it is thrown up mechanically mixed with the steam.

That the circulation of water may be good, the passages for the descending currents should be large and clean, and so far protected from the action of the heat, as to prevent its interference with the descending current. To this end, short cylinder boilers with the fire under them should have one end at least protected from the heat, and the brick-work on the sides should come down so low as to give the water a good start down before the action of the heat takes effect upon it; and the tubes or flues of such boilers must be kept so far from the shells, and so far apart, as to leave good room for these currents. There is no danger but that water will rise, but great care is needed in the provision for the descending currents. Many boilers now in use would give better results, because of the improvement in the force of the descending currents, that would be made by the removal of quite a percentage of their tube surface, or by the covering up of quite a percentage of their shells.

Designers of boilers have been misled, in regard to the amount of surface that could be put into a shell with advantage, by the locomotive boiler; but the packing of tubes into a locomotive, the front end of which is comparatively cool, and with no heat acting upon any part of its shell to interfere with the descending currents, cannot be safely followed in boilers which have most of the lower parts of their shells exposed to the direct action of the fire and heated gases. And undoubtedly many locomotives, even, have more tubes in them than is for their good.

Explosions, from a heating up of the water to a temperature above that due the pressure, can be prevented by the use of boilers having a strength in accordance with the rule on page eighty-five, and by the use of safety-plugs so constructed that their strength will be so far reduced as to be broken by a very low pressure, before there is such an elevation of temperature as to endanger the boiler, — the breaking of the safety-plug causing the gradual escape of the overheated water, and reduction of the temperature of the fire, without any such agitation as to lead to the explosive giving off of steam by the overheated water. It will also be well in all cases, when a boiler has been standing quiet for any length of time, to avoid the sudden opening of a valve, so as to cause a sudden reduction of pressure and agitation of the water; * and also to avoid taking in feed-water at such a time so fast as to cause much agitation in the water, or in such quantities as to introduce a great amount of air with it. And when the fires have been banked up for a considerable length of time, I think it will be well, after they are cleaned out, but before they get to burning up so as to impart much heat to the boiler, to either blow off a little water, let a little steam escape, or take in a little feed-water, so as to cause a very little agitation of the water, in order that enough little particles of solid matter be put in motion, to prevent the overheating of the water, before the fire gets strong enough to cause the rising of these little particles of solid matter, when they have all been precipitated far from the furnace.

But, in my opinion, perfect immunity from the danger of explosion, from the overheating of the

---

* And, in fact, it is well always to avoid the sudden opening of any valve of such a size as to cause a rapid reduction of pressure.

water, can only be secured by the employment of boilers having a strength as great as has been before indicated, and of such a construction that all their surfaces can be reached so that they can be kept clean; and with such provision for the circulation of the water, that safety-plugs may be used so weak as to be broken by an overheating much below the point of danger to the boiler. The doing of these things will be attended with such advantages as to well repay the doing of them, even without taking into consideration the fact of freedom from the danger of explosion.

Also From Merchant Books

www.ingramcontent.com/pod-product-compliance
Lightning Source LLC
Chambersburg PA
CBHW051416200326
41520CB00023B/7259